D0721996

# THE PRACTICE
# OF BAYESIAN
# ANALYSIS

# THE PRACTICE
# OF BAYESIAN ANALYSIS

Edited by

**Simon French**
*Professor of Informatics, School of Informatics,*
*The University of Manchester, UK*

and

**Jim Q Smith**
*Professor of Statistics, University of Warwick, UK*

A member of the Hodder Headline Group
LONDON • SYDNEY • AUCKLAND

Copublished in North, Central and South America by
John Wiley & Sons, Inc., New York • Toronto

First published in Great Britain in 1997 by
Arnold, a member of the Hodder Headline Group,
338 Euston Road, London NW1 3BH

Copublished in North, Central and South America by
John Wiley & Sons, Inc., 605 Third Avenue,
New York, NY 10158-0012

*British Library Cataloguing in Publication Data*
A catalogue record for this book is available from the British Library

*Library of Congress Cataloging-in-Publication Data*
A catalog record for this book is available from the Library of Congress

ISBN: 0 340 66240 9
ISBN: 0 471 19477 8 (Wiley)

Typeset by Mathematical Composition Setters Ltd, Salisbury, Wiltshire
Printed in Great Britain by J.W. Arrowsmith Ltd., Bristol

# Contents

# List of contributors

*Tim Bedford*
Delft University of Technology, Faculty of Mathematics & Computer
Science, P.O. Box 5031, 2600 GA Delft, The Netherlands

*May Chan*
Clinical Information Science Unit, 26 Clarendon Road, Leeds LS2 9NZ,
UK

*Susan E Clamp*
Clinical Information Science Unit, 26 Clarendon Road, Leeds LS2 9NZ,
UK
s.clamp@leeds.ac.uk

*Roger M Cooke*
Delft University of Technology, Faculty of Mathematics & Computer
Science, P.O. Box 5031, 2600 GA Delft, The Netherlands
r.m.cooke@twi.tudelft.nl

*F Tim de Dombal*
(deceased)

*Joan Dorrepaal*
Delft University of Technology, Faculty of Mathematics & Computer
Science, P.O. Box 5031, 2600 GA Delft, The Netherlands

*Malcolm Farrow*
School of Computing and Information Systems, University of Sunderland,
St Peter's Way, Sunderland SR6 0DD, UK
m.farrow@sunderland.ac.uk

*Simon French*
School of Informatics, University of Manchester, Oxford Road, Manchester
M13 9PL, UK
simon.french@man.ac.uk

*Michael Goldstein*
Department of Mathematical Sciences, University of Durham, Science
Laboratories, South Road, Durham DH1 3LE, UK
michael.goldstein@durham.ac.uk

*Miles T Harrison*
School of Computer Studies, University of Leeds, Leeds LS2 9JT, UK

*David Ríos Insua*
Departamento de Inteligencia Artificial, Facultad de Informática,
Universidad Politécnica de Madrid, Campus de Montegancedo, s/n,
Boadilla del Monte, 28660 Madrid, Spain
drios@fi.upm.es

*Karen E Jenni*
Department of Engineering and Public Policy, Carnegie Mellon University,
Pittsburgh, PA 15213, USA

*Matthijs Kok*
HKV Consultants, PO Box 2120, 8203 AC Lelystad, The Netherlands
m.kok@hkv.nl

*Miley W Merkhofer*
Applied Decision Analysis, Inc., 2710 Sand Hill Road, Menlo Park, CA
94025, USA

*Jan M van Noortwijk*
HKV Consultants, PO Box 2120, 8203 AC Lelystad, The Netherlands
j.m.van.noortwijk@hkv.nl

*Tony O'Hagan*
Department of Mathematics, The University of Nottingham, University
Park, Nottingham NG7 2RD, UK
aoh@maths.nott.ac.uk

*Catriona M Queen*
Institute of Mathematics and Statistics, University of Kent, Canterbury,
Kent CT2 7NF, UK
c.m.queen@ukc.ac.uk

*David C Ranyard*
School of Computer Studies, University of Leeds, Leeds LS2 9JT, UK
dcr@scs.leeds.ac.uk

*Peter J Regan*
Strategic Decisions Group, Two International Place, 20th Floor, Boston,
MA 02110, USA
coitregan@aol.com

*Stephen Senn*
Department of Statistical Science, University College, Gower Street,
London WC1E 6BT, UK
stephens@public-health.ucl.ac.uk

*Jim Q Smith*
Department of Statistics, University of Warwick, Coventry CV4 7AL, UK
stran@snow.csv.warwick.ac.uk

*Takis Spiropoulos*
Business School, University of Hertfordshire, Hertford Campus, Mangrove
Road, Hertford SG13 8QF, UK
e.spiropoulos@herts.ac.uk

*Carol Williams*
JK Research Associates, Inc., 86 Gold Hill Road, Breckenridge, CO 80424,
USA

# Preface

The Bayesian approach to statistics and decision analysis has now established itself as a powerful and very practical tool to support the non-statistical, but otherwise technical, modeller. Recent advances in computational techniques have released its potential to an extent where even very complicated scenarios are amenable to a Bayesian analysis. The technical advances which have made this possible are now well documented, so that there is a need to present and emphasize some of the modelling issues which must be addressed in Bayesian analyses, whether conducted by the statistical community or others.

The purpose of this book is twofold. Firstly, it has been written to advertise the advantages a Bayesian analysis can bring. New statistical and decision models can be tailored to the unique beliefs, values and needs of the user, and the implications of the data she collects can be analysed with reference to this underlying structure. The second purpose is to provide an analyst wanting to apply the Bayesian methodology with a resource of examples and techniques, so that she may be better informed about the scope of opportunities and the pitfalls inherent in such an analysis. In short, the emphasis of this book is on the practice of Bayesian model building and analysis as a response to the needs of the user. Obviously, this short compilation only scratches the surface of this subject. We hope, however, that it provides enough examples and pointers to be a useful contribution for helping to develop communication channels between the Bayesian practitioner and her client.

We would like to thank all the authors who have contributed papers to this compilation. Sadly, one contributor, Tim de Dombal, passed away as the volume was in preparation. His co-authors kindly brought the paper into its final form – we all hope that it is presented in a way he would have wanted.

As well as thanking all the contributors, we would like to thank all those involved in the production of this volume and in particular Nicki Dennis for her gentle but persistent prompting which ensured this project was completed almost to schedule.

Jim Q Smith, Simon French
November 1996

# 1 Bayesian analysis

## S French, J Q Smith

## 1.1 The Bayesian paradigm

The Bayesian approach to statistical inference and decision analysis may be described in many ways, some alas so simplistic that its subtlety and power can easily be misunderstood. Perhaps the most simplistic of these, yet in a way the most accurate, is the mathematical one.

Stated simply, a decision maker has to choose a single action $a \in A$ from a set of possible actions. The consequence $c(a, \theta) \in C$, which might be multidimensional or *multi-attributed*, of her[1] choice depends on both which action she takes and an unknown state of nature $\theta$, which she knows to lie in the set $\Theta$. Before choosing the action, she may observe an outcome $X = x$ of an experiment, which depends on the unknown state $\theta$. Specifically, the observation $X$ is drawn from a distribution $P_X(. \mid \theta)$.

The decision maker's objectives are encoded in a real-valued *loss function*:

$$\ell(a, \theta) = -u[c(a, \theta)] \tag{1.1}$$

which measures the loss or negative utility of the consequence $c(a, \theta)$ to her. Thus her problem is:

> The Statistical Decision Problem – Observe $X = x$ and then decide on an action $d(x) = a$ using this information to minimize, in some sense, $\ell[d(x), \theta]$.

The Bayesian solution is to encode her prior knowledge of $\theta \in \Theta$ through subjective probabilities via her *prior distribution* $P_\Theta(.)$. Her prior knowledge is updated in the light of the observation $X = x$ using Bayes' theorem to give her *posterior distribution* $P_\Theta(. \mid \theta)$:

$$P_\Theta(\theta \mid x) \underset{\theta}{\propto} P_X(x \mid \theta) P_\Theta(\theta) \tag{1.2}$$

---

[1] Decision makers will be referred to in the feminine or plural, while statisticians and decision analysts will be referred to in the masculine. Apart from the niceties of political correctness, the use of distinct pronouns helps present their roles without confusion or needless repetition of 'decision maker' and 'statistician'.

Or, in words, *her posterior distribution is proportional to the likelihood multiplied by her prior distribution*, where the *likelihood* is $P_X(x \mid \theta)$, considered as a function of $\theta$ for fixed $X = x$, the observation. The constant of proportionality may be found simply, since her posterior distribution $P_\Theta(. \mid \theta)$ must integrate or sum to 1 with respect to $\theta$.

Then, with the objective of minimizing her *posterior expected loss*, she should choose the action $d(x) = a$ as

$$d(x) = \arg_a\min\{E_\theta[\ell(a, \theta) \mid x]\} \tag{1.3}$$

Equations (1.1)–(1.3) describe the *subjective expected utility (SEU) model* formulated by, *inter alia*, Savage (1954). There is a confusing difference of perspectives between Bayesian statisticians and Bayesian decision analysts in that the former work in terms of losses, while the latter tend to work in terms of utilities. Operationally, the only difference is whether one includes a minus sign as in eq. (1.1) and minimizes or maximizes in eq. (1.3). Either way, one refers to SEU theory.

Before we move on to other entry points to the Bayesian approach, a couple of points in the above should be noted. Firstly, the decision maker's knowledge is modelled in a number of ways. The obvious – and most controversial to non-Bayesians – is the use of probability to represent her degrees of belief about $\theta$. This introduces issues of subjectivity, which we discuss below. But the definitions of the parameter, action and consequence spaces along with the mapping $c: A \times \Theta \rightarrow C$ also model her knowledge, her view of the world and her assumptions about it. A major task in any analysis is to identify the issues before her and to define these spaces and mapping appropriately so that the mathematical formulation is seen as appropriate and sufficiently detailed – or *requisite* – to capture all the perceived important relationships and objectives that need to be modelled. We pick up this point in Section 1.3.3.

This characterization frames the Bayesian approach in terms of *decision making*. Statistics, many would argue, is about *inference*. Some might go further and argue that inference – the abstraction of information from data – should not be biased by the possible uses to which that knowledge might subsequently be put. We shall refrain from entering into that debate. Here we simply note that Bayesian analysis can focus on inference without any reference to decision. The scientist, as we shall call her in such contexts rather than a decision maker, simply expresses her initial beliefs about the likely value of some parameter $\theta \in \Theta$ though her prior probability distribution and updates this in the light of the observed data $X = x$ using Bayes' theorem as in eq. (1.2). There is no need to define any action or consequence space, or any mapping $c: A \times \Theta \rightarrow C$, nor to elicit a utility function to model her preferences. Moreover, the report of her analysis is provided by a statement of her posterior distribution. There is no need to define families of estimators, types of confidence intervals or hypothesis tests. Indeed, if one tries to take a Bayesian view of these entities, one is drawn back towards a decision theoretic view of statistics (Berger, 1985; French, 1998; Robert,

1994). Thus, for a Bayesian who focuses solely on inference, the methodology would appear to be remarkably straightforward: construct $P_\Theta(. \mid \theta)$ via eq. (1.2). But that characterization does ignore the rather complex problem of how she should report and communicate an often multidimensional probability distribution and what features in this distribution to consider (*see* Section 1.5).

All this is, of course, a thumbnail sketch of a whole raft of theory which we are taking for granted in this collection of readings. We refer the reader unfamiliar with Bayesian analysis to Berry (1996), Lee (1989) and Smith (1997).

Taking a less mathematical view of the Bayesian school, we may characterize it by the explicit introduction of judgement and prior beliefs into the model. There is no drive to obtain a goal of 'scientific objectivity' in the sense of analyses which are completely independent of their participants, i.e. independent of the decision maker or scientist and analyst. Instead, a Bayesian analysis seeks to model explicitly relevant judgements of the participants. It is openly subjective. By so being, it seeks to allow other decision makers and scientists to remove the participants' judgements from the analysis and replace them with their own so that they may see what they would have decided or inferred in the same circumstances with the same data. Thus for a Bayesian, scientific knowledge emerges not from objective analyses but from consensus, when a majority of scientists agree that each starting from her prior beliefs is led to the same inference on the basis of the data.

The willingness to model judgement immediately introduces the question of how one *should* model judgement, particularly beliefs or expressions of uncertainty. The Bayesians have addressed this since the mid-eighteenth century and by a variety of routes, axiomatic and heuristic, have been led to the modelling of belief by probabilities and preferences by utilities as in the SEU model described in eqs (1.1)–(1.3) above. The justifications of this approach are legion (see references at the end of this chapter). With the exception of a few minor subtleties, all justifications lead to the same model. Moreover, it is a model which is essentially very simple, yet applies to an enormous variety of circumstances. There is no need to invent *ad hoc* methods to deal with specific circumstances, although we should admit that this simplicity of methodological outlook is bought at a cost of computational difficulty. Until the advent of cheap modern computing power with high quality graphics and visualization techniques, it was often impossible to evaluate the posterior through an application of Bayes' theorem and to calculate, much less minimize, an expected utility.

Fig. 1.1 guides us towards a further perspective on the Bayesian approach. It begins by structuring the problem to separate issues of uncertainty, belief and knowledge from issues of preference or value judgement. The former are modelled by probabilities, the latter by utilities, and they are treated separately for much of the analysis, only being recombined when the expected utilities of the possible actions are formed. This separation is the

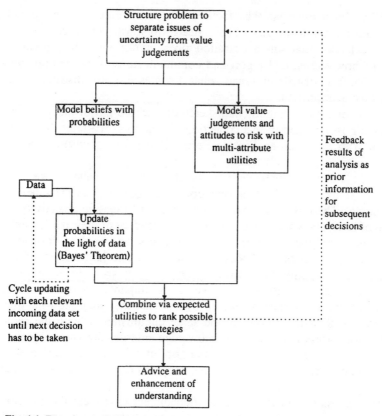

**Fig. 1.1** Bayesian analysis

first stage in a series of separations which mean that Bayesian analysis is truly *analytic*, i.e. it breaks a problem down into component parts, analyses each of these in turn and then combines them into a synthesis which balances each part. At the highest level of separation, which is shown in the figure, issues relating to belief are separated from those related to preferences. At lower levels of the analysis, concepts of exchangeability or of probabilistic independence and dependence are used to decompose joint probability distributions into products of marginal and conditional distributions. Similarly, using preferential and utility independence multi-attribute utility functions can be decomposed into various simpler forms.

There are three other points we would note from Fig. 1.1. Firstly, Bayesian analysis allows cycling at two points. As more and more data arrive for analysis, the application of Bayes' theorem to update probabilities may be repeated until the time comes to make a decision. Moreover, the output of one analysis can feed into a subsequent analysis. Secondly, note the lowest box in the figure: 'Advice and enhancement of understanding'. As we shall emphasize in Section 1.3.3, the direct output of an analysis should not be a decision or an inference. Rather, it is *understanding* which helps the decision maker to decide or the scientist to infer. Analyses should support decision making and inferences, not supplant them.

The third point is one that we shall note again and again in this collection of readings: the importance of *modelling*. Fig. 1.1 requires us both to 'model beliefs with probabilities' and to 'model value judgements and attitudes to risk with multi-attribute utilities'. Moreover, the structuring of the problem 'to separate issues of uncertainty from value judgements' provides a very specific model of a decision problem (French, 1998). It is this modelling that provides much of the 'art' of Bayesian analysis, allowing the analyst to help the decision makers and scientists to explore subtleties in reaching their understanding.

## 1.2   The actors

### 1.2.1   Decision makers, scientists, experts and analysts

Subjective analyses, such as the Bayesian, directly involve people – and those people have distinct roles to play. The basic axiomatic presentations of the theory identify just one role: that of the decision maker. She is assumed to be her own knowledge expert, to express her own preferences, perform her own analyses and make her own decisions. In practice, this seldom happens. Rather a group of decision makers, e.g. a management board, aware that they have a decision to make, will ask an analyst or perhaps a consulting company to consult scientists or subject experts to gain relevant information, then to formulate the problem, gather and analyse data and advise on the decision. A similar tale can be told in the case of inference. Science seldom advances through individuals working alone. Generally, there is a team effort, involving scientists, their peer reviewers and statistical analysts.

Thus, many will be involved in, or provide input to, an analysis, but only very occasionally will any single person be responsible for all its parts. Note that some of the participants will provide beliefs and knowledge, some will provide value judgements, and some both – and a few, the analysts, will provide no content input, but rather process expertise. The analysts provide skills and expertise in elicitation, modelling and analysis, but their beliefs and value judgements in relation to the context should not be involved. We begin with the case of decision making and then turn to inference.

The *decision makers* own the problem, although that does not mean that they know exactly what the problem is. Very often, decision analyses begin when the decision makers are aware of a raft of issues that need resolving, but are unable even to describe what such a resolution might look like. For instance, a major company might be concerned at the possible advent of a single European currency unit and the effect that this would have on its long-term trading position. Yet the company might have no clear idea of what choices are before them, and even less of an idea about how to evaluate those choices. Thus they seek help from an analyst.

The *(decision) analyst* helps the decision makers to discuss, explore and formulate the problem. The art of problem formulation is well discussed in operational research (see, e.g., White, 1985; Rosenhead, 1989; Keeney, 1992; Daellenbach, 1994), but perhaps less well discussed in the statistical literature, although we should mention Kadane (1996) and Gelman *et al.* (1995). Once the problem is formulated, the analyst elicits detailed belief and value judgements, collects or acquires the necessary data, performs the calculations and presents the results back to the decision makers. Note that, in performing these tasks, the analyst needs not only technical skills, but also communication and, often, group process skills – see Eden and Radford (1990) for a discussion of some of the latter issues and the techniques and skills required. We emphasize that the role of the analyst is to help the decision makers to an enhanced understanding so that through this understanding they, not the analyst, may make a better informed decision.

Both in formulating the problem and in eliciting quantities to include in the analysis, the analyst may need to consult more widely than the decision makers. An *expert* is anyone consulted by the analyst to provide beliefs relating to the unknown state of the world. Experts provide information relevant to the probability models, i.e. the input relates to the left-hand branch of Fig. 1.1. The inclusion of expert judgement into an analysis is a complex issue and one that has in the past led to some confusion and less than clear thinking on the part of Bayesians. We address it here, partly because of its importance *per se* and partly because doing so explains our current emphasis on the need to be clear on people's role in an analysis. In the simple SEU model, the decision maker encodes her beliefs as probabilities. Bayesian analysis requires that a posterior distribution be developed. So it is natural when consulting an expert to ask him for his probabilities and use these in the analysis. However, the axioms underpinning SEU theory assumes that they are *her* probabilities: it is a truly subjective theory. There are good reasons why the decision maker might not wish to adopt the expert's probabilities for her own.

Firstly, experts may have good context knowledge, but may lack skills in encoding that knowledge through probabilities. Fig. 1.2 indicates one issue here. If an expert repeatedly gives many probability judgements on a series of events, one might expect that on those occasions he announces a probability of, say, 0.4, the events concerned happen 40 per cent of the time. In such cases, the expert is said to be (*frequency*) *calibrated*. Alas, few experts are. Their calibration curves are not the 45° line, but depart from it, typically as shown in Fig. 1.2, although often with greater magnitude. Moreover, the calibration curve of an expert may be context-specific. For instance, a financial expert may be quite differently calibrated in matters concerning foreign currency markets than in European stocks. Thus the decision maker would be wise to allow for such issues. In addition, she herself may well have prior information relevant to the decision: few decision makers are totally ignorant of the contexts in which they operate.

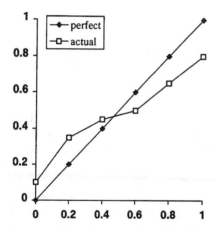

**Fig. 1.2** A calibration curve for an expert

Thus there is a need to combine her prior opinions with those of the expert. The Bayesian view is that this should be done via Bayes' theorem. Asking an expert for information is no different from performing an experiment: both provide data (see French, 1985 for details).

In the above, we have been making a further dubious though ill-discussed assumption. Namely, if a decision maker wants advice on the likelihood of some state or event, the expert should be asked to announce a probability. The axiomatic basis of SEU leaves no doubt that a decision maker should encode her uncertainty through probabilities – but it says nothing about how an expert should. Communicating uncertainty is a very different task from using uncertainty measures in calculations. It may be that experts find formalisms other than probability a more natural, and perhaps a more precise, vehicle for communication. Without training, few are good probability assessors, and even with training there is no guarantee that any individual expert can assess probabilities well and thus communicate his uncertainty accurately Other methods for encoding uncertainty may serve the task of communication better. As we have indicated, this has not been well discussed in the literature: a recent paper by Ofir and Reddy (1996) indicated some early comparative experiments.

The decision makers may also consult *stakeholders*. Stakeholders share the consequences of the decision with the decision maker, although they do not take part in the process of deciding. Obvious examples of stakeholders are shareholders in a company or members of the public in relation to decisions on environmental policy. If stakeholders' interests are to be represented in the analysis, they take on a role similar to that of experts: namely, they provide information on *preferences* rather than *beliefs*. Their role relates to the right-hand branch of Fig. 1.1. However, in other senses, stakeholders are very different from experts. Whereas it seems rational for the decision maker and analyst to assimilate the expert's opinions in a manner which adjusts for calibration issues, it would seem very inequitable, or at best maternalistic, for her to adjust the preferences of her stakeholders. We shall discuss these issues further in the next section.

If we turn now to inference rather than decision making, we find a slightly different set of roles. Firstly, the decision maker is replaced by the *scientist*. She is the person who formulates a model – often referred to as a scientific hypothesis – to be tested empirically or a quantity to be determined. She designs an experiment or series of experiments, ideally in consultation with a (*statistical*) *analyst*. As above, the analyst helps to formulate the model statistically, structure the analysis and communicate the results. At this point, a third set of people enter the fray: *peer reviewers* (Smith 1996 refers to them as *auditors*). Science advances, in our Bayesian view, when the results and ideas of one group of scientists are read, checked and adopted by a majority of the whole scientific community. Thus a Bayesian analysis must allow the peer reviewers (i.e. other scientists) to extract the prior opinions of the scientist who conducted the experiment and substitute their own to see if they too reach the same inference. In O'Hagan (Chapter 8) there is a role of audit within a business analysis.

### 1.2.2   Individuals and groups

The perceptive reader will note that we have been rather lax in our use of singular and plural in discussing the roles of the decision maker(s) and others above. The knowledgeable reader will know that we have committed a multitude of sins in doing so. Now is the time to begin seeking atonement.

The SEU model strictly only applies to a single decision maker. Yet, as we have noted, most decisions are made by groups on behalf of stakeholders and involve expert judgement. Surely, therefore, we should develop a theory of group SEU and base Bayesian methodology on that. Alas, this is not possible. In a celebrated theorem, Kenneth Arrow has shown that Savage's theory cannot be extended to apply to groups of decision makers – see French (1985, 1986) for a discussion of this result. So how do we reconcile our belief in the appropriateness of the Bayesian approach with this uncomfortable result?

The answer lies in the subjectivism of the Bayesian school. The methods are explicitly subjective and subjectivism focuses on the individual. Groups do not make decisions: individuals do. Groups are social processes which translate the decisions of individuals – their votes – into a single action. It is a habit of language that we speak of this group process as 'decision making', but it is very different from the deliberations which lead individuals to make a choice.

From this perspective it is easy to appreciate how Bayesian ideas support groups of decision makers. The analyses:

- help each individual in the group to understand the decision from their own viewpoint and, hence, to choose, were the decision theirs alone;

- allow group members to communicate their own views of the problem to other members of the group and, in turn, understand the perspectives of other group members;

- through sensitivity analysis (see Section 1.5.3) identify the real issues of contention and avoid sterile debate about disputed input judgements that do not lead to a difference in the final decision.

Group issues do matter in Bayesian decision analyses, but not in the sense of constructing a single agreed probability distribution and group utility leading to a single group expected utility ranking. Rather, Bayesian analyses for groups seek to improve communication, cooperation and mutual understanding.

Our arguments have focused on group issues relating to the decision makers. When groups of experts or groups of stakeholders are involved, there are further issues to be considered. Firstly, in taking the view that experts provide data, we should not consider them independent data sources. Two experts inevitably share some common background; for instance, it is likely that they have had a similar education, that they have read many of the same papers and seen some of the same data sets. This means that their judgements are likely to be correlated[2] and this needs to recognized in the analysis. Two experts saying much the same thing for much the same reason does not provide twice the information. However, the preferences expressed by two stakeholders are also likely to be correlated. Equity demands that two stakeholders who express the same preference are given twice the weight. These are, of course, rather simplistic statements, but they further emphasize the need to think clearly about the distinct roles of the participants in an analysis.

## 1.3   Elicitation and modelling issues

Let us move on to other matters. In order to apply the SEU model, the analyst needs to elicit probabilities and utilities from the decision maker or scientist. In doing so, he will discover that she does not express intuitive judgements that obey the assumptions inherent in SEU theory. Judgements of uncertainty may be biased or poorly calibrated as in Fig. 1.2. Moreover, the decision maker is unlikely to provide a description of the world which may be structured easily and quickly into a model suitable for analysis. Thus the analyst must work with her to develop her perception of the problem context and to build a model which reflects this. All of the case studies to follow address these issues in detail and we shall not steal their thunder. But we do provide some background.

### 1.3.1   Psychological issues

There have been many experimental studies to discover how intuitive judgements fit with the assumptions of the SEU model. Most have discovered that the fit can be very poor indeed. Unaided human judgement is susceptible to many biases. We are well aware that in using the word 'bias' in this context, we are implicitly assuming that the SEU model is correct or ideal in some

---

[2] Indeed, their judgements may be correlated with those of the decision maker for similar reasons (see French, 1985).

way. We make that assumption without further comment here – discussion may be found in French (1986, 1998). So how does unaided human judgement depart from the SEU model? There are many ways: we have already noted calibration (or the lack of calibration) of probability judgements. Here we note only a few further effects to give an impression of the issues.

- *Ignoring base rates and overweighing concrete evidence.* These are *very* non-Bayesian effects! Essentially, when presented with data, individuals forget all they knew before and do not combine the data with their prior knowledge. Thus, for example, on learning that a medical test has indicated that they have a rare disease, they may assume that they have that illness. Yet, if the test can give false positives, the posterior probability given by an application of Bayes' theorem illness may be very low.

- *Misconceptions of 'randomness'.* In a sequence of 10 tosses of a coin, HHHHHTTTTT is equally as likely as HHTTTHTHHT to occur, yet individuals behave as if the latter is the more likely. In making judgements of likelihood, individuals often expect events to 'look' random. In a similar manner, they may forget effects such as regression to the mean.

- *Availability.* In judging probabilities, subjects often recall series of similar circumstances and ask themselves in what frequency the event of interest occurred. They may often generate very atypical sets of 'similar circumstances' and hence produce biased estimates. For instance, if asked for the probability that a person will fall ill with a neurological disease in the next year, an individual may think of what proportion of her friends have such a disease. Since such diseases are often memorably horrific, she may tend to overestimate the incidence by forgetting quite how many friends she has who do not have the disease.

- *Anchoring.* People tend to 'anchor' on the first number they hear or estimate in a given context. Thus an analyst worth his salt would never ask: 'Is the probability of rain tomorrow greater or less than 80 per cent?' because the answer will be nearer 80 per cent than a more neutral question would have elicited.

- *Framing.* The manner in which a question is framed can affect the answer dramatically. For instance, in a classic experiment conducted originally by Kahneman and Tversky and repeated by hundreds of lecturers with their students since, subjects were given the following situation and asked to choose between programmes A and B:

Imagine that you are a public health official and that an influenza epidemic is expected. Without any action it is expected to lead to 600 deaths. However, there are two vaccination programmes that you may implement: programme A would use a well established vaccine which would save 200 lives, programme B would use a new vaccine which might be effective. There is a one-third chance that it would save all 600 lives and a two-thirds chance of it saving none of them.

About three-quarters of all subjects prefer A. Other subjects have been presented with the same scenario and asked to choose between programmes C and D:

Programme C would use a well established vaccine which would lead to 400 of the population dying, programme D would use a new vaccine which might be effective. There is a one-third chance of no deaths and a two-thirds chance of 600 deaths.

About three-quarters of these prefer D. But, of course, the choice in each case is the same: programmes A and C are identical in their outcomes, as are programmes B and D. In other words, it matters whether a question is framed positively or negatively.

There are many other 'biases' which have been encountered in behavioural studies: we leave the reader to find descriptions of them in the literature (see, e.g. Kahneman *et al.*, 1982). In working with his client, the analyst must remember that she may be susceptible to such biases and take care to explore her intuitive statements to help her reflect upon and remove any possible bias.

### 1.3.2  Descriptive and normative modelling in prescriptive analysis

The early discussions of Bayesian methodology made a strong distinction between *normative* models and *descriptive* models. Normative models suggest how people *should* make inferences and decisions, descriptive models describe how they *do*. This dichotomy was at the heart of many of the debates about the value of Bayesian analyses over the past 40 to 50 years. The central issue, as we have just noted, is that very seldom do people make inferences and decisions in accordance with the axioms underlying the SEU model. So, argued the opponents, Bayesian ideas are clearly unacceptable and inappropriate to the majority of scientists and decision makers. No, retorted the Bayesians, just because people's *unguided* inferences and decisions do not correspond with the assumptions of rationality underpinning the theory does not mean that they should not. People make mistakes. Just as human mental arithmetic is an imperfect activity when compared with laws of arithmetic as enshrined in the foundations of mathematics, so are the inference and decision processes. And just as they need calculators and computers to guide their arithmetic, so they need Bayesian analyses to guide their inferences and decisions. We shall not rehearse these arguments further, but note that recently the Bayesians have slightly softened their stance through the introduction of the concept of *prescriptive analyses* halfway between the normative and descriptive. Roughly, a prescriptive approach to statistical inference and decision analysis is indicated in Fig. 1.3.

Prescriptive analyses guide scientists and decision makers towards an inference or decision by providing a normative model which captures both aspects of the problem before them and aspects of their beliefs and preferences. This model or, on occasion, family of models helps them

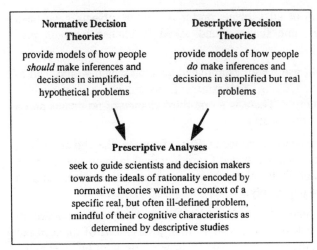

**Fig. 1.3** Prescriptive analyses

towards a greater understanding both of the problem and of themselves. Through this understanding, they make the inference or decision. However, in communicating with the scientists and decision makers and in eliciting their judgements, the analyst needs to understand how they make inferences and decisions unaided. Thus both normative and descriptive models contribute to the analysis. For further discussion, see, *inter alia*, Bell *et al.* (1988) and French (1998).

### 1.3.3    Requisite modelling

Prescriptive analyses guide the evolution of decision makers' perceptions. During prescriptive analyses their perceptions change and their understanding grows. It is the purpose of the analysis that they should. Thus it is vital to see the modelling process involved in representing their perceptions as creative, dynamic and cyclic. The decision makers' beliefs and preferences are assessed and modelled, and the models are explored, leading to insights and a revision of their judgements, and thence revision of the models used. The process cycles until no new insights are found. Phillips (1984) described this evolution, referring to the process as *requisite modelling*, the final model being requisite or sufficient for the inference or decision faced. Too often a static view is taken of decision analysis, in which the beliefs and preferences – indeed all the judgements of the decision makers – are taken as fixed and immutable from the outset of the analysis. It is not a view which we shall adopt. The purpose of the analysis is to guide this evolution towards a set of judgements that are more consistent and rational, as well as better understood than they were previously.

We have all faced complex issues such as buying a house or choosing a career. How many of us have begun with a clear and complete understanding of our feelings? Rather, we start out muddled, clear on our thoughts in some respects, e.g. 'I do want a house with a garden' or 'I do not want to have a career which involves substantial living away from home', but unclear on others: 'What are the chances that I will need a spare bedroom?'; 'Can I be sure of passing the professional examinations?'. As we think about such problems, we clarify our thoughts. Our beliefs and preferences evolve. The role of an analyst is to help his client evolve her feelings and her perceptions. One mechanism which he uses is to help her break her problem down into a number of separate issues, focus on each in turn, reflect on related information that she may have but not realize, gather further information if required and then pull all the pieces together into an overall, balanced view: an enhanced understanding. Throughout, he works at developing the *process* while she provides the *content*.

### 1.3.4 Forms of model

Earlier we remarked that Bayesian analysis was truly analytic – it separates issues into their components and looks at each of these separately before reassembling them into an overall perspective. This is achieved through the structure provided by statistical and decision models. We have in mind here model forms such as influence diagrams, decision trees, statistical hierarchical models, belief nets and attribute hierarchies.

Influence diagrams and decision trees give graphical representations of different perspectives of the overall decision problem (*see* Figs 1.4 and 1.5 for examples). Influence diagrams show belief or knowledge relationships, whereas decision trees show temporal relationships. These perspectives are complementary, a fact often overlooked by some of the discussions in the literature in which the relative virtues of the two representations have been extolled. Fig. 1.5 clearly shows the temporal relationships between the decisions and the outcome of the test results, but hides, for instance, the fact that the test result will depend on whether the equipment is predisposed to fail. In more complex problems, decision trees also hide a multitude of conditional dependencies. On the other hand, Fig. 1.4 shows the dependencies clearly, but hides the temporal relationships. Indeed, to represent this 'asymmetric' tree as an influence diagram requires the introduction of some modelling tricks such

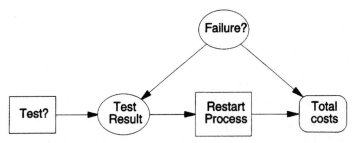

**Fig. 1.4** Influence diagram for a decision on whether to test equipment further before returning it to service after maintenance

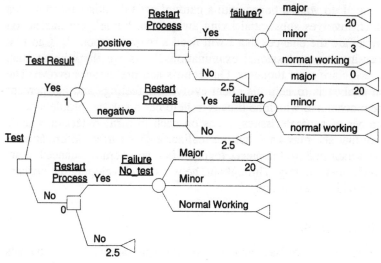

**Fig. 1.5** Decision tree for the decision on whether to test equipment further before returning it to service after maintenance

as null tests. This can confuse the decision maker about some aspects of her problem, whereas the decision tree maps out the future for her.

Both decision trees and influence diagrams provide perspectives on the whole decision problem; other modelling techniques focus on the belief or the preference aspects separately. Statistical hierarchical models, for instance, represent the decision maker's beliefs about the unknown state $\theta$ by building a sequence of parameters and hyperparameters, related to each other probabilistically. An example is given in Fig. 1.6 (for further details, see, for example, Lindley and Smith, 1972; Box and Taio, 1973; Bernardo and Smith, 1994; O'Hagan, 1994; Gelman *et al.*, 1995).

More recently, much more elaborate models have been studied, often using graphical methods and belief nets (Oliver and Smith, 1990; Marshall and Oliver, 1995; Lauritzen, 1996). Essentially, these generalize the linear

---

$P_X\left(\cdot|\theta\right)$ might be structured as:

$$\theta_1 \sim P_{\theta_1}\left(\cdot|\theta_0\right)$$

$$\theta_2 \sim P_{\theta_2}\left(\cdot|\theta_1\right)$$

$$X \sim P_X\left(\cdot|\theta_2\right)$$

Thus, on writing $\theta = \left(\theta_0,\theta_1,\theta_2\right)$ and expressing in terms of densities:

$$p_X\left(x|\theta\right) = p_X\left(x|\theta_2\right)p_{\theta_2}\left(\theta_2|\theta_1\right)p_{\theta_1}\left(\theta_1|\theta_0\right)$$

**Fig. 1.6** A simple hierarchical model

ordering of dependence implicit in hierarchical models to a partial one. Such models are illustrated in the case studies to follow (Queen, Chapter 10; Ranyard and Smith, Chapter 11).

On the preference side of things, attribute hierarchies provide structure over which the decision maker may explore and articulate her preferences. We use the term *attribute* to mean one of the dimensions along which she assesses the consequences of her decision. Thus, 'cost', 'market share' or 'loss of life' might be attributes. A *criterion* or *objective* refers to an attribute and a direction of preference. Thus, 'minimize cost', 'maximize market share' and 'minimize loss of life' are criteria. Some attributes are objective in that they correspond to physical measurements, while others are more subjective, requiring judgement in their definition, e.g. equity of treatment. It is conventional to organize the attributes involved in analysis into hierarchies. This offers many cognitive advantages and also helps structure decision analyses (Keeney, 1992). Fig. 1.7 provides an example of

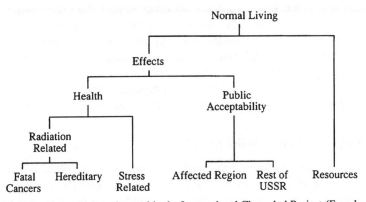

**Fig. 1.7** Attribute hierarchy used in the International Chernobyl Project (French, 1996)

an attribute hierarchy used to explore the value judgements underpinning decisions made after the Chernobyl accident.

### 1.3.5   Model building

As we have noted, the client does not normally approach the analyst with a neat model which is ready formulated. The process of model building is one of the key stages in an analysis, one which provides much insight and understanding. We focus on building a statistical model for inference: the discussion goes through *mutatis mutandis* for preference modelling and for the task of assembling the components into a full decision model (Keeney, 1992).

Traditional approaches set up a statistical model much as an 'Aunt Sally'. It may contain premises, such as the equality of certain parameters, which are believed by the client to be wrong and a traditional analysis would provide evidence to support this falsity. Such analyses have their place, but do not provide creative input to the user. In contrast, the advantage of the subjective approach to modelling is that it can reflect a train of thought through the introduction of latent and unobservable variables. A good Bayesian analysis starts with what Wemuth and Lauritzen (1990) call a research hypothesis, namely, a probabilistic coding of what the client currently believes about the dependence structure of the variables in her problem. The analysis of the data then suggests to the client the way in which she should adjust those beliefs. Because the evidence is used systematically to evolve the client's belief structure, it is creative and evocative, allowing her to discover new relationships and interpretations, which dovetail into the framework in which she naturally works.

However a model is elicited, the analyst must ensure that the model faithfully represents the client's beliefs. He should also ensure that she understands the logical implications of her belief statements as indicated by the model, firstly to help her understand the context further, and secondly as a consistency check on the elicitation. In practice, this elicitation process often provokes resistance initially, if for no other reason than the client's increasing awareness of her beliefs and her accountability should they be erroneous. Nevertheless, in due course the opportunities for realism and flexibility in such a model, which can be adapted and adjusted to a changing environment, usually overcome any such initial hesitation.

The elicitation process can be divided conceptually into two phases, although in practice it may cycle between them rather than proceed linearly.

### 1.3.6   Eliciting a qualitative belief model

The client is encouraged to discuss the concerns and issues which have led her to seek an analyst's help. From this discussion, important uncertain quantities begin to emerge, along with an understanding of her objectives. The importance of these quantities derives from their perceived role in achieving her objectives. These quantities often form a crude hierarchical structure. At the first level are quantities which impact directly on her objectives. These first level quantities will be affected by circumstances in a second level of variables, and so on. In most non-trivial hierarchical models there are at least three levels of variables:

- first level quantities, analogous to dependent variables in statistical analyses;

- second level quantities, corresponding to states of the process;

- third level quantities, corresponding to independent variables.

Often there are more than three levels.

Having made an initial identification of the quantities which are key to a client's problem, we invoke the Clarity Test (Howard, 1990). Quantities which are sufficiently well defined to be realizable at some future time – at least in principle – are called *target variables*. For instance, a target variable might be the rainfall in mm at a given site during a given period. The client is encouraged to reflect on those quantities which are too nebulous to be defined as target variables. Sometimes she recognizes a lack of clarity in her thinking and redefines them sufficiently tightly for them to become target variables; otherwise they are set aside to take an evocative rather than a central role in the analysis, setting the scene to help to assess the existence and strength of relationships between target variables.

Next – although again we emphasize that the modelling process is very far from a linear one – the analyst works with his client to obtain a provisional specification as to how variables relate to each other. This might be done:

- graphically, using chain graphs or belief nets (Lauritzen, 1996);

- by eliciting various exchangeable or partially exchangeable structures (see, e.g., O'Hagan, 1994);

- by the specification of function equations with error, like those appearing in state space forecasting (see, e.g., West and Harrison, 1989);

- by a combination of all three of the above.

These structures are discussed with the client and modified until they seem to be requisite. This process is described for influence diagrams in Smith (1994, 1997). The analyst now has a provisional structure for the model which reflects the client's view of her world sufficiently well to continue with the analysis, although, of course, subsequent insights may lead to further revisions in the model's structure.

### 1.3.7   Quantifying a client's belief model

In some simple discrete models, it is possible to move from the qualitative phase of elicitation to direct assessment of the probabilities required. For the most part, however, it will be necessary to select families of continuous distributions over the uncertain quantities. There will always be a degree of arbitrariness at this stage, although this can be offset to some extent by using characterizations of specific distributions (Johnston and Kotz, 1969, 1970a, 1970b, 1972; Lauritzen, 1982). For example, if the distribution of independent variables $X_1$ and $X_2$ is such that $(X_1 - X_2)$ and $(X_1 + X_2)$ are also independent, then $(X_1, X_2)$ must be Gaussian. Checking the validity of such statements can justify the use of a particular family of distributions.

Whatever the choice of a family of distributions, it is important that they are selected so that:

1. they respect the dependence structure elicited in the qualitative phase;

2. they are rich enough to capture most of the dependencies that the client wants to explore (or that the analyst believes that she will need to explore);

3. they are sufficiently simple to work with that analyses can be completed in time.

Next, probability or parameter values need to be elicited. Ideally, these should be obtained by asking the client questions about realizable quantities of which she has experience. For example: 'On giving treatment A to 100 patients who exhibit characteristics and symptoms B, how many would you expect to have their temperature reduce to below 37.8 °C within 2 hours?' From replies to questions such as these, the analyst's task is to construct the prior probability distributions over the less tangible quantities in the model. Great care is needed because it is easy for the analyst to introduce unintentional biases into the client's statements, e.g. by poor framing of the questions (see Section 1.3.1 above).

The analyst should take care to reflect the model's implications back to his client as a check. Often, apparently trivial assumptions made by the analyst to expedite the analysis have implications which the client finds unacceptable, even preposterous, if she is made aware of them. This constant reflection of the model back to the client is an essential part of producing requisite analyses.

There are many ways in which possible difficulties may be avoided. The first is to forego the possibility of specifying a full distributional model and instead elicit a set of means and covariances associated with selected functions of these quantities (see, e.g., Farrow *et al.*, Chapter 4). A second, and in our opinion somewhat less satisfactory, method is to assign 'vague' prior distributions to higher level parameters in the hierarchy. An apparently compelling way of doing this is via the concept of invariance (Berger, 1985; cf. the discussion of characterizations above). For instance, if adding an arbitrary number to an uncertain quantity would not affect her probability that its value was less than zero, there would be location invariance which would enable the analyst to build the model a little more precisely. However, unlike the use of characterization results, the ensuing analyses may be non-Bayesian in that incoherence may result, e.g. any natural group invariant structure for a covariance matrix of two or more Gaussian variables will introduce probability statements which would lead the client to take a sure loss on certain bets (Eaton and Sudderth, 1993). Even when such anomalies do not occur, the implications of vague priors on statements about observable quantities in the model would not usually bear close scrutiny. Thus we advise that vague priors are only used as a last resort.

## 1.4   Bayesian analysis

For a book on the practice of Bayesian analysis, we are going to say remarkably little on the analysis of the model in the introduction. Indeed, few of the following papers focus on this aspect of the Bayesian approach to inference and decision. The reason for this is that Bayesian analysis essentially requires us to do two things:

• apply Bayes' theorem to construct the posterior distribution in the light of data;

• maximize the posterior expected utility in order to guide choice.

How these tasks might be achieved is essentially irrelevant to the import of the analysis. As Bayesians we may be delighted that the constraint of being bound by the algebra of conjugate distributions has been removed by advances in computation. Nowadays Bayes' theorem can be applied numerically or via Monte Carlo simulations. There are ingenious algorithms for propagating evidence and maximizing expected utility on belief nets and influence diagrams. There are now many software tools available, which protect the analyst from a need to be a *technical* probability and computational expert – but not from the need to understand the ideas *conceptually*! These advances in analysis do not concern the client *per se*. She is interested in the insight and understanding brought by the analysis, not in the mechanism by which it does so. Thus we shall take the process of analysis for granted and move on.

## 1.5 Reporting and presenting results

The purpose of the analysis being to enhance the client's insight and understanding, it is vital that its import is communicated to her effectively. Many recent advances in computer graphics have come to a Bayesian's aid in reporting results. It is now possible to visualize and explore distributions in ways that were not possible 10–15 years ago. See, for example, Cooke *et al.* (Chapter 2) and remember that a black and white, static illustration is rather poor at conveying the power of modem dynamic graphics with colour, rendering and shadow which allow the user to change viewpoints and vary the projection studied. Instead of discussing the technology by which this is achieved, we provide somewhat more general remarks on the process of reporting and presenting results in order to ensure that the analysis is requisite.

### 1.5.1 Facilitation

In the past the lack of portable, instantly available computer power has meant that the analyst has usually 'taken the problem away from his client, solved it and returned to present her with a solution'. This has tended to focus attention on the technical skills of the analyst. Today, it is often possible to formulate, analyse and solve the problem in the presence of the client. Certainly, it is possible to explore variations in the structure of the model(s) in front of the client. This has meant that the analyst's role has developed into someone who facilitates the communication between his client and the model as much as that of a technician. Today's analyst requires 'people' skills even more than in the past. He – or at least one of his team – must be willing to adopt and communicate in his client's language.

There are many benefits in doing so. The immediacy of the analysis and the clarity with which it can be communicated mean that clients 'buy into' the analysis more fully. They feel that it is *their* analysis, and hence are more ready to champion and implement its conclusions. Again, we refer to Eden and Radford (1990) for a discussion of these issues.

### 1.5.2    The power of the model's structure to enhance understanding

Often the reporting emphasizes the output of the analysis – and there are obvious and appropriate reasons why it should do so. However, the analyst is unwise to forget the power of the structure of the model to enhance the client's understanding. Remember that the client usually approaches an analyst for help with an awareness that she has a problem or set of issues to tackle, but a lack of clarity as to precisely what they are. Eliciting and structuring a model bring much insight and this should be reflected upon and confirmed in the reporting. Despite the subtlety and sophistication of much of the technical analysis, it may be something rather more 'trivial' to the analyst that brings the client the greatest understanding. For instance, many decision analysts have reported that simply displaying the contingencies of a decision by means of a decision tree has been sufficient to enable the client to see her solution: the analysis of the tree was the icing of on the cake (see, e.g., French, 1988).

### 1.5.3    Sensitivity analysis

Early presentations of the Bayesian approach, taking a very normative line, downplayed the role of sensitivity analysis, or even denied that it had one (see, e.g., Savage, 1954). From the modern perspective of prescriptive support of inference and decision, sensitivity analyses – or robustness studies, as they are often termed within statistics – have a more central role in Bayesian studies.

Technically, sensitivity analysis is the investigation of the effect of changes in the data input to a model on its output. Sensitivity analysis enables the analyst and decision maker to explore whether there is a clearly preferred alternative, or whether there are several strongly competing alternatives. Sensitivity analyses enable the identification of dominated and near-dominated alternatives, i.e. those alternatives which would never be preferred whatever weights were assigned to criteria. In doing so, they focus attention on those parts of the model and analysis which drive the conclusion, rather than those aspects which seem important at the outset.

At the cognitive or individual level, sensitivity analysis is the process of interactively exploring the effects of changes in the inputs to and structure of a model in order to learn about the problem and about the client's values and priorities. The viewpoint here is the growth in understanding of each individual decision maker of her perceptions (see especially Keeney, 1992 and Phillips, 1984).

Finally, from the social or group viewpoint, sensitivity analyses can help groups of decision makers focus on their *real* differences. Often a heated discussion of issues on which individuals fundamentally disagree can be completely defused by a sensitivity analysis which shows that, despite differences on the appropriate weights or scores, the decision should be the same. Sensitivity analyses also help individual group members communicate their views to the other members of the group (cf. our remarks in Section 1.2.2).

## 1.6 Concluding remarks

We close this introduction, in a sense, as we began by arguing that the strength of the Bayesian approach to inference and decision is far stronger than simply its normative basis. We believe in Bayesian analysis for the following reasons:

- *Axiomatic basis*. An undoubted strength of the Bayesian approach is the axiomatic basis of the SEU model. Many have investigated the assumptions underpinning the model and they are both well understood and persuasive. Moreover, when related axiom systems, e.g. of the propagation of evidence, are constructed, they are found to fit Bayesian approaches (Oliver and Smith, 1990; Jensen, 1996).

- *Lack of counterexamples*. Not all principles of rationality are encoded as axioms. It is possible that some have been overlooked in formulating the axiom system and are not implied by it. Were this the case, it would be possible to construct examples in which unacceptable behaviour was supported. We are unaware of such examples and their absence strengthens the Bayesian case. That is, of course, not to suggest that counterexamples will not arise, only that those which have arisen have been explained by a careful consideration of the theory. For instance, there were many anomalies in the early Bayesian approaches to the use of expert judgement, but the introduction of appropriate correlation structures overcame them (French, 1985).

- *Feasibility*. However sound the Bayesian approach is theoretically, to be applied it must be feasible. Recent computational advances have ensured that it is. The studies that follow are an eloquent testament to this.

- *Transparent to users*. Any analysis must be meaningful to its users. They must be able to understand:

  - what the inputs mean
  - what calculations have been made
  - what the results mean.

  We stressed many points in the discussion of the elicitation of a model to indicate that the model and its analysis should be meaningful to the client. Again we point to the studies that follow to demonstrate that this is both possible and natural.

- *Robust*. The sensitivity of an analysis to variations in the inputs should be identifiable and understood. Again, computational advances have allowed fuller sensitivity analyses to be conducted.

- *Compatible with a wider philosophy.* The models and analysis should 'fit' with the decision makers' wider view of the world. This point has often been overlooked, demonstrating a certain arrogance on the part of analysts, Bayesian and non Bayesian alike. Any analysis must answer the questions asked by the clients in their language and terms. The studies, in particular Ranyard and Smith (Chapter 11) and de Dombal *et al.* (Chapter 3), demonstrate that at least Bayesian approaches can do as well in this respect as any other approach; but it is a sensitive issue that each analyst must tackle afresh each time he begins an analysis.

## 1.7   Further reading

Introductions to Bayesian statistics may be found in Berry (1996), Lee (1989) and Smith (1997). More advanced presentations and current surveys of the literature are available in Berger (1985), Bernardo and Smith (1994), Berry *et al.* (1996), French (1998), Gelman *et al.* (1995), O'Hagan (1994) and Robert (1994). There are many other texts: two classic texts are Box and Taio (1973) and DeGroot (1970). Introductions to Bayesian decision analysis may be found in Clemen (1990), French (1988), Smith (1997) and Watson and Buede (1987). More advanced presentations are provided by, *inter alia*, French (1986), Keeney and Raiffa (1976) and Von Winterfeldt and Edwards (1986). For an understanding of the *process* of modelling and analysis, we particularly recommend Bell *et al.* (1988), Eden and Radford (1990), Daellenbach (1994), Gelman *et al.* (1995), Kadane (1996), Keeney (1992) and Kleindorfer *et al.* (1993). An excellent summary of issues relating to the incorporation of expert judgement into an analysis may be found in Cooke (1991), albeit not as fully a Bayesian view as in French (1985). There are many excellent collection of readings on behavioural decision making, i.e. on the results of experiments into unaided human judgement. We refer to Kahneman *et al.* (1982) particularly. A more recent text is provided by Wright and Ayton (1994). However, as Beach *et al.* (1987) note, negative results in the sense of failing to fit the SEU model tend to receive more attention than positive ones: there are citation biases as well as behavioural ones.

## References

Beach, L.R., Christensen-Szalanski, J. and Barnes, V. 1987: Assessing human judgement: has it been done, can it be done, should it be done? In Wright, G. and Ayton, P. (eds) *Judgemental forecasting*. Chichester: John Wiley and Sons.

Bell, D.E., Raiffa, H. and Tversky, A. (eds) 1988: *Decision making*. Cambridge: Cambridge University Press.

Berger, J.O. 1985: *Statistical decision theory and Bayesian analysis*, 2nd edn. New York: Springer-Verlag.

Bernardo, J.M. and Smith, A.F.M. 1994: *Bayesian theory*. Chichester: John Wiley & Sons.

Berry, D.A. 1996: *Statistics: a Bayesian perspective*. Belmont: Duxbury Press.

Berry, D.A., Chaloner, K.M. and Geweke, J.K. (eds) 1996: *Bayesian analysis in statistics and econometrics: essays in honour of Arnold Zellner*. New York: John Wiley and Sons.

Box, G.E.P. and Taio, G.C. 1973: *Bayesian inference in statistical analysis*. New York: Addison Wesley.

Clemen, R. 1990: *Making hard decisions*. Belmont: Duxbury Press.

Cooke, R.M. 1991: *Experts in uncertainty: opinion and subjective probability in science*. Oxford: Oxford University Press.

Daellenbach, H.G. 1994: *Systems and decision making*. Chichester: John Wiley and Sons.

DeGroot, M.H. 1970: *Optimal statistical decisions*. New York: McGraw-Hill.

Eaton, M.L. and Sudderth W.D. 1993: Prediction in a multivariate normal setting: coherence and incoherence. *Sankya* **A55**, 481–95.

Eden, C. and Radford J. (eds) 1990: *Tackling strategic problems*. London: Sage.

French, S. 1985: Group consensus probability distributions: a critical survey. In Bernardo, J.M., DeGroot, M.H., Lindley, D.V. and Smith, A.F.M. (eds) *Bayesian statistics 2*. Amsterdam: North Holland, 183–201.

French, S. 1986: *Decision theory: an introduction to the mathematics of rationality*. Chichester: Ellis Horwood.

French, S. (ed.) 1988: *Readings in decision analysis*. London: Chapman and Hall.

French, S. 1996: Multi-attribute decision support in the event of a nuclear accident. *Journal of Multi Criteria Decision Analysis* **5**, 39–57.

French, S. 1998: *Statistical decision theory*. London: Arnold.

Gelman, A., Carlin, J.B., Stern, H.S. and Rubin, D.B. 1995: *Bayesian data analysis*. London: Chapman and Hall.

Howard, R. 1990: From influence to relevance to knowledge. In Oliver, R.M. and Smith, J.Q. (eds) *Influence diagrams, belief nets and decision analysis*. Chichester: John Wiley, 3–23.

Jensen, F.V. 1996: *Bayesian networks*. London: University College Press.

Johnson, N.L. and Kotz, S. 1969: *Distributions in statistics: discrete distributions*. New York: John Wiley and Sons.

Johnson, N.L. and Kotz, S. 1970a: *Distributions in statistics: continuous univariate distributions 1*. New York: John Wiley and Sons.

Johnson, N.L. and Kotz, S. 1970b: *Distributions in statistics: continuous univariate distributions 2*. New York: John Wiley and Sons.

Johnson, N.L. and Kotz, S. 1972: *Distributions in statistics: continuous multivariate distributions*. New York: John Wiley and Sons.

Kadane, J.B. (ed.) 1996: *Bayesian methods and ethics in a clinical trial design*. New York: John Wiley and Sons.

Kahneman, D., Slovic, P. and Tversky, A. (eds) 1982: *Judgement under uncertainty*. Cambridge: Cambridge University Press.

Keeney, R.L. 1992: *Value-focused thinking: a path to creative decision making*. Cambridge, MA: Harvard University Press.

Keeney, R.L. and Raiffa, H. 1976: *Decisions with multiple objectives: preferences and value trade offs*. New York: John Wiley and Sons.

Kleindorfer, P.R., Kunreuther, H.C. and Schoemaker, P.J. 1993: *Decision sciences*. Cambridge: Cambridge University Press.

Lauritzen, S.L. 1982: *Statistical models as extremal families*. Aalborg: Aalborg University Press.

Lauritzen, S.L. 1996: *Graphical models*. Oxford: Oxford University Press.

Lee, P. 1989: *Bayesian statistics: an introduction.* Oxford: Oxford University Press.

Lindley, D.V. and Smith, A.F.M. 1972: 'Bayes' estimates for the linear model'. *Journal of the Royal Statistical Society* **B34**, 1–41.

Marshall, K.T. and Oliver, R.M. 1995: *Decision making and forecasting.* New York: McGraw-Hill.

Ofir, C. and Reddy, S.K. 1996: Measurement errors in probability judgements. *Management Science* **42**, 1308–25.

O'Hagan, A. 1994: *Kendall's advanced theory of statistics, Vol. 2B. Bayesian inference.* London: Edward Arnold.

Oliver, R.M. and Smith, J.Q. (eds) 1990: *Influence diagrams, belief nets and decision analysis.* Chichester: John Wiley and Sons.

Phillips, L.D. 1984: A theory of requisite decision models. *Acta Psychologica* **56**, 29–48.

Robert, C. 1994: *The Bayesian case.* Berlin: Springer Verlag.

Rosenhead, J. (ed.) 1989: *Rational analysis for a problematic world.* Chichester: John Wiley and Sons.

Savage, L.J. 1954: *The foundations of statistics*, 1st edn. New York: John Wiley and Sons (2nd edn, 1972). New York: Dover.

Smith, J.Q. 1994: Decision influence diagrams and their uses. In Rios, S. (ed.) *Decision theory and decision analysis: trends and challenges.* Boston, MA: Kluwer, 33–51.

Smith, J.Q. 1996: Plausible Bayesian games. In Bernardo, J.M., Berger, J.O., Dawid, A.P. and Smith A.F.M. (eds) *Bayesian statistics 5.* Oxford: Oxford University Press.

Smith, J.Q. 1997: *Decision analysis: a Bayesian approach*, 2nd edn. London: Chapman and Hall.

Von Winterfeldt, D. and Edwards, W. 1986: *Decision analysis and behavioral research.* Cambridge: Cambridge University Press.

Watson, S.R. and Buede, D.M. 1987: *Decision synthesis: the principles and practice of decision analysis.* Cambridge: Cambridge University Press.

Wemuth, N. and Lauritzen, S.L. 1990. On substantive research hypotheses, conditional independence graphs and graphical chain models. *Journal of the Royal Statistical Society* **B52**, 21–50.

West, M. and Harrison, P.J. 1989: *Bayesian forecasting and dynamic models.* New York: Springer-Verlag.

White, D.J. 1985: *Operational research.* Chichester: John Wiley and Sons.

Wright, G. and Ayton, P. (eds) 1994: *Subjective probability.* Chichester: John Wiley and Sons.

# 2 Mathematical review of Swedish Bayesian methodology for nuclear plant reliability data bases[1]

R M Cooke, J Dorrepaal, T Bedford

## 2.1 Introduction

This article is a shortened version of a report commissioned by the Swedish Nuclear Inspectorate SKI (Cooke *et al.*, 1995) to review the Bayesian data processing methodology (Pörn, 1990) implemented in the T-Book. The authors found it convenient to distinguish generic issues involved with the design of reliability data bases (RDBs) from specific issues relating to SKI's Bayesian implementation. Part 1 therefore focuses on these generic issues, and part 2 addresses the specific issues.

Part 1 contains a summary of reliability data base design concepts, introducing the decisions which must be taken in designing data collection and processing methodology. This discussion is quite summary in nature. A more complete discussion is found in Cooke *et al.* (1993) and in a recent special issue of *Reliability Engineering and System Safety*.

The main conclusion is that current data processing methods do not satisfactorily account for competing failure modes. The most dramatic example of this concerns the failure modes *critical failure* and *preventive maintenance* (degraded or incipient failure); in fact, the current data processing methodologies effectively assume that the rate of occurrence of critical failures is unaffected by the rate of occurrence of preventive maintenance.

The objectives of part 2 are to:

- review the mathematical model underlying the T-Book methodology (Pörn, 1990);

[1] This work was performed for SKI under contract 94367, 14.2-940997. Many of the ideas described here were developed in research for the European Space Agency.

- develop independent computer implementations to serve as an independent check on the results reported in Pörn (1990);

- Identify key assumptions of the T-Book methodology and explore the sensitivity of the model to these assumptions.

The overall conclusion of part 2 is that the T-Book methodology represents a significant advance over current RDB methods. The mathematical model is sound, its assumptions are plausible, if not inevitable, and the independent computer implementation produced good agreement with the results stated in Pörn (1990). Nevertheless, this method does inherit some of the generic problems in RDB design identified in part 1, and further research in this field would be worthwhile.


# Generic issues (part 1)

## 2.2   Reliability data base (RDB) users

Modern reliability data banks (RDBs) are intended to serve at least three types of users:

1. the component designer interested in optimizing component performance;

2. the maintenance engineer interested in measuring and optimizing maintenance performance;

3. the risk/reliability analyst wishing to predict reliability of complex systems in which the component operates.

The desire to serve this variety of users has driven RDB designers to distinguish up to 10 distinct failure modes, often categorized as *critical*, *degraded* and *incipient* failures. This development can be traced back to the IEEE format of 1984, and is reflected in modern publicly available data bases such as those of the Institute of Chemical Process Safety (ICPS), OREDA and EIREDA.

The failure modes are distinct ways in which a component can terminate a service sojourn. RDB designers are thus faced with the analysis of *competing risk* data. Distinct causes compete, as it were, to kill the component. For each service sojourn, only one cause will be seen and recorded, namely, the one which actually 'kills' the component. We do not see other causes which might have killed the component at a later time.

The theory of competing risk was developed in the 1970s. Paradigm applications were drawn from the fields of biostatistics and analysis of mortality tables. This theory assumes that the competing risks operate independently. The main mathematical result states that the (possibly time-dependent) failure rates associated with competing risks can be uniquely identified from competing risk data *if* the risk are independent. This mathematical fact provides the basis for all processing methodologies currently employed in RDB design.

The question is whether the independence assumptions underlying the current methods are appropriate. It is well to realize that a 'small' error in methodology can propagate to a large error in applications, when this error is repeated on multiple applications.

## 2.3 Example

A simple example illustrates the problem of competing failure modes. A reliability engineer buys a new car and collects reliability life data. He logs the times at which his car is taken in to repair a breakdown (failure). He maintains his car fastidiously: every Sunday morning he repairs anything which is not functioning perfectly even though no breakdown has occurred. These 'non-critical repairs' are logged as preventive maintenance. His neighbour, who never works on his car, asks the engineer how reliable the new car is. What should he say?

If he uses the standard methods employed in RDBs, he will estimate the (by assumption constant) rate of occurrence of breakdowns (failure rate) via the 'total time on test statistic': the number of failures divided by the total time in operation. The total time on test statistic is the *observed rate of failure*, i.e. the rate at which failures befall the fastidious engineer in spite of his preventive maintenance. It is also something else, and herein resides much confusion. *If* the preventive maintenance were performed randomly, then the total time on test statistic *also* estimates the rate of failures when no maintenance is performed. This latter is called the *naked failure rate*.

Now, what does the neighbour want to know? He does not want to know the rate of failure if he spends every Sunday on his car, as he will not do this. He wants to know the rate of failures when no maintenance is performed. Another neighbour might perform modest maintenance; he would then adapt the naked failure rate to his own situation to predict the rate of failure. We return to the distinction of observed and naked failure rates in Section 2.6.

The standard method assumes that preventive repairs and failures are statistically independent. If this assumption were true, our reliability engineer would be a very bad reliability engineer indeed, for he would be performing ineffective maintenance on his car – failures would be just as frequent if he did nothing! Preventive maintenance is supposed to *prevent* failures while losing as little useful service time as possible. This entails that preventive maintenance should be highly correlated to failure. Ideally, the car is preventively maintained at time $t$ if and only if it *would have* failed shortly after $t$.

A reliability engineer performing good maintenance and using the standard methods will come up with an overoptimistic estimate of the naked failure rate of his car. This is because the standard methods do not properly account for the fact that maintenance is preventing failures from occurring.

At the same time, he would come with an overpessimistic assessment of his maintenance performance. In other words, the standard methods will analyse 'good maintenance on a bad component' as 'bad maintenance on a good component'. Indeed, the rate of occurrence of failures, as analysed by standard methods, is not affected by maintenance!

## 2.4  Raw data

The raw data which forms input to a RDB processing methodology may be described as 'component socket time histories'. A component socket is a functional position in a system occupied by one component during one service sojourn. A service sojourn begins when a new or repaired component goes on line, and terminates when the component is removed for any reason whatever. A typical format for component socket time histories is shown in Table 2.1. The repair fields would give the calendar hours and man hours that the component was in repair, the time at which the component socket

**Table 2.1** Format for component socket time histories

| Plant | Socket | Time/date | Repair fields | Failure fields |
|-------|--------|-----------|---------------|----------------|
| • | • | • | • | • |
| • | • | • | • | • |
| B1 | 311v1 | 9:00/2-2-78 | How long | Cause, mechanism, effect... |
| B1 | 311v1 | 14:00/4-6-84 | How long | Cause, mechanism, effect... |
| B1 | 311v1 | 17:00/5-8-94 | How long | Cause, mechanism, effect... |
| • | • | • | • | • |
| B1 | 311v2 | 8:00/3-6-75 | How long | Cause, mechanism, effect... |
| • | • | • | • | • |
| • | • | • | • | • |
| O1 | 311v1 | .................... | | |
| • | • | • | • | • |
| • | • | • | • | • |

was replenished, as well as other relevant information, e.g. whether the repair was done on site. The failure fields would describe the method of detection, the failure cause, failure mechanism and functional consequence of failure.

A RDB processing methodology must take such raw data and convert it into information which the RDB users need. We may distinguish three steps in the processing:

1. Screening – remove duplications, resolve ambiguities, fill in blank fields, if possible;

2. Pooling – cluster failure fields data into distinct cells, group sockets and plants;

3. Computing – convert grouped data into numerical representations.

In this report, we focus on the third step.

## 2.5   Computational models

We assume that the plant–socket combinations of interest have been defined, and that the failure cells have been defined. We may regard these cells as distinct failure modes. Each service sojourn terminates in one and only one such failure mode. We assume further that a socket is replenished with a component which is as good as new.

Suppose three plant–sockets are pooled together and that two distinct failure modes, C and P, are distinguished (later these are associated with corrective and preventive maintenance). Then data for this pool could be represented graphically as in Fig. 2.1.

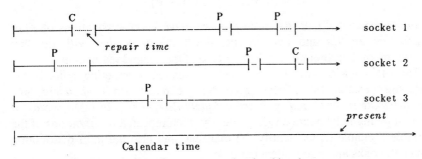

**Fig. 2.1** Graphic representation of component socket time histories

Such pooling is usually done on the assumption that the pooled sockets are independent and that the probabilities of failure modes C and P are the same at each socket and do not depend on calendar time. Note that observation is terminated at the 'present' and that a residual service time is logged at each socket.

In this case we can put all the data on a common time axis by concatenating the time bars and grouping the residual time in service at the end. In the graph in Fig. 2.2, time has been rescaled to fit on the page and repair times

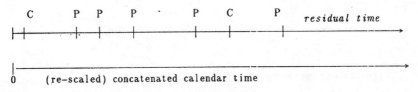

**Fig. 2.2** Concatenated component socket histories

have been removed. This is the format for data on which the mathematical processing models operate. The data may now be regarded as a renewal process with concatenated time as parameter. There are two styles of modelling this process, namely as a competing risk process and as a coloured Poisson process.

### 2.5.1 Competing risk

From the point of view of competing risk, the failure modes C and P are competing to kill the component. Every time the component is killed, it is replaced (immediately, in the above picture) by a 'good as new' component. Hence we never observe C and P together; we always see whichever happens first. C and P are modelled as distinct random variables of which we observe the smaller. In other words, we observe independent and identically distributed copies of the the random variable $Y$ ($I\{C<P\}$ is the indicator function of the event $\{C<P\}$; it tells us whether the minimum is an instance of C or of P):

$$Y = \{\min(C, P), \quad I\{C<P\}\}$$

In the concatenated calendar time plot, we have seven realizations of $Y$, of which two are Cs and five are Ps; the value associated with the first realization of C is about one-half the value of the second realization.

The main mathematical result for independent competing risks is the following (Tsiatis, 1975; Peterson, 1977): *if C and P are independent, then the distributions of C and of P are uniquely determined.* In other words, if the competing risks operate independently, then repeated observations of the variable $Y$ enable us to identify the marginal and also the joint distribution (since the risks are independent) uniquely.

On the other hand, if the competing risks are *not* independent, then it is *not possible*, in general, to identify the (marginal) distributions from the competing risk data.

### 2.5.2 Coloured Poisson process

In this style, we think of the labels 'C' and 'P' as colourings of events. We considered the 'uncoloured process' by writing U for C and for P (*see* Fig. 2.3). Assume that the uncoloured process is a Poisson process with

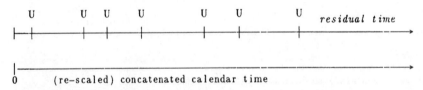

**Fig. 2.3** Uncoloured Poisson process

intensity $\gamma$.[2] Assume that each event is coloured by flipping an independent coin with $P(\text{heads}) = p$. If heads turns up, then the event is coloured 'C', otherwise it is coloured 'P'. The 'colouring theorem' says that the resulting processes of events C and P are independent Poisson processes with intensities $\gamma p$ and $\gamma(1 - p)$ respectively (Kingman, 1993).

The methodology incorporated in the T-Book (Pörn, 1990) assumes that critical failures form a Poisson process. The information from other failure modes (e.g. non-critical failures, preventive maintenance) is treated in a manner consistent with the assumption that the entire process, including all failure modes, is a coloured Poisson process. This means, intuitively, that the 'P' coloured points contain no information about the 'C' coloured points.

### 2.5.3 Relation between competing risk and coloured Poisson processes

The relationship between these two types of models is easily given (see Cooke, 1996). The coloured Poisson process is mathematically equivalent with the independent exponential competing risk model. If C has constant failure rate $\lambda_C$, and P has constant failure rate $\lambda_P$, independent of C, then $\min(C, P)$ is exponentially distributed with failure rate $\lambda_C + \lambda_P$. This is the distribution of inter-arrival times in the uncoloured Poisson model. $\gamma = (\lambda_C + \lambda_P)$ and $\lambda_C = p\gamma$, $\lambda_P = (1 - p)\gamma$ with $p = \lambda_C/(\lambda_C + \lambda_P)$.

In spite of this equivalence, the language of competing risk is conceptually richer and somewhat more amenable to describing non-constant failure rates and non-independent risks than is the language of coloured processes.

## 2.6 Naked versus observed failure rates

In Section 2.3 we distinguished informally between the 'observed failure rate' and the 'naked failure rate'. We can now define these notions a bit more precisely, and formulate the relation between them.

We first recall the notion of a survival function and failure rate. The survival functions for C and P are

$$S_C(t) = \text{Prob}(C > t)$$

$$S_P(t) = \text{Prob}(P > t)$$

Note that $S_C(t)$ cannot in general be estimated from the competing risk data. The failure rate for C is given by

$$r_C(t) = \text{Prob}\{C \in (t, t + \delta) | C > t\}/dt = \frac{- dS_C(t)/dt}{S_C(t)}$$

In a competing risk context, $r_C$ is the marginal failure rate of C. It is the failure rate which we *would* observe for C, if we could observe *all* Cs, even

---

[2] The probability of $n$ uncoloured events in a time interval of length $T$ is $(\gamma T)^n \exp(-\gamma T)/n!$; and the expected number of events in this interval is $\gamma T$.

those Cs which were larger than the competing P. This is what we called the *naked failure rate* in Section 2.3. In speaking of a naked failure rate, we assume that it is meaningful to speak of the value which C *would* have if P had not occurred first. In many situations, this is indeed the case. Hence, we can speak of the time at which a valve *would* have failed due to leaking seals, even if it fails earlier due to cracked membrane; and we speak of the time at which a component would have failed critically even though it was preventively maintained before critical failure.

The observed failure rate is defined as:

$$\text{obr}_C(t)dt = \text{Prob}\{C \in (t, t + \delta) \text{ AND } C < P \mid C > t \text{ AND } P > t\}$$

The fundamental fact regarding the naked and observed failure rates is related to the result quoted in the previous section (see Cooke, 1996): if C and P are independent then

$$r_C(t) = \text{obr}_C(t) \qquad r_P(t) = \text{obr}_P(t)$$

In other words, if the competing risks are independent, then the two failure rates coincide; if the risks are not independent, then the two failure rates do not necessarily coincide.

It follows from the above that, if the competing risks are not independent, then we cannot identify the naked failure rates from the competing risk data; there may be many naked failure rates which are consistent with the same (infinitely many) observations of the variable Y.

What use is the naked failure rate notion in this context? Well, first of all it is essential to realize what the observed failure rate is *not*! More importantly, we can explore additional modelling assumptions which would determine the naked failure rate, and we can bound possible values of the naked failure rates from the observed data. These are examples of qualitative or exploratory data analysis.

### 2.6.1 Rates of occurrence

The rate of occurrence of a renewal process at time $t$ is defined as

$$\text{Rate of occurrence}(t) = (d/dt)[\text{expected number of events in } [0, t]]$$

For exponential renewal processes, the rate of occurrence and the failure rate coincide, but in general they are distinct concepts. It is useful to compare the *observed* and the *naked* rates of occurrence for exponential competing risks. We reproduce in Fig. 2.4 the concatenated time history of Fig. 2.2, showing an interval of length $I$ which is spanned by the (concatenated) observations. The *(empirical) observed rate of occurrence* of critical failure 'C', in interval $I$ is $2/I$. This is the 'total time on test statistic' (TTT):

$$\text{TTT} = \text{no. of critical failures/total observation time}$$

With every value of 'P' there is associated some value of 'C' which *would* have been observed if the P had not occurred first. The *empirical naked rate of occurrence* of critical failure is the number of observed and unobserved

Cs in *I*, divided by *I*. Obviously, we cannot determine this from the observations alone, but we can say:

$$1/I \le \textit{empirical naked rate of occurrence of } C \le 7/I$$

The lower bound $1/I$ derives from the fact that the history in Fig 2.4 begins with a 'C', otherwise the lower bound would be zero. The upper bound would

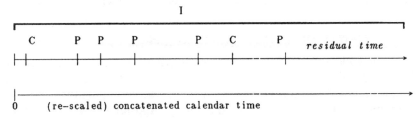

**Fig. 2.4** Concatenated component socket histories

be obtained if each unobserved 'C' were to occur immediately after the observed 'P'. Failures would then be intercepted just before they happen. This would correspond to the ideal preventive maintenance policy: obtaining maximal component life with minimum failures. Slightly less effective maintenance results in 'P's occurring before the 'C's, thus censoring useful component life. If the first 'P' censored useful life greater than 'I – time to first C', then the empirical naked rate of occurrence would be $1/I$.

The meaning of the independence assumption now becomes clear: if C and P are independent competing risks, then the observed rate of occurrence is equal to the naked rate of occurrence, $2/I$.

## 2.7   Conclusions (part 1)

We cannot estimate naked failure rates or rates of occurrence from data without additional assumptions. How can an RDB support the users identified in Section 2.2? There are basically two ways.

First, recall that the independence assumption is *always* compatible with observed failure data, although possibly *incorrect*. If independence does not hold, then some form of dependence must hold. However, the various simple dependence models are *not always* compatible with the failure data. Hence an RDB should present the component socket histories in such a way as to enable an evaluation of different models for the interaction (dependence) of competing failure modes.

Second, the previous section showed that bounds can be placed on the naked failure rates based on observed data. With appropriate use of graphical tools, the amount of uncertainty in naked failure rates can be presented in such a way as to distinguish sampling uncertainty from *identifiability* uncertainty.

These methods have been implemented in various pilot projects (Cooke *et al.*, 1993; Cooke, 1996; Paulsen *et al.*, 1996) to which the reader is referred for details and examples.

# Mathematical review (part 2)

Part 2 is specifically directed to the data processing methodology developed by SKI for application in the T-Book. Section 2.8 reviews the mathematical model and its underlying assumptions; Section 2.9 studies the inference model and defines verification tasks to be performed with independent code. Section 2.10 describes two ways in which this model has been coded, and Section 2.11 gives the results of the verification tasks. Section 2.12 gathers conclusions.

## 2.8   Review of the mathematical model

The mathematical model developed in Pörn (1990) is designed to solve the following problem: how should data from similar plants be combined? The solution proposed in Pörn (1990) is, in our view, conceptually superior to earlier proposals found in the literature (Apostolakis *et al.*, 1980; Apostolakis, 1982; Kaplan, 1983; Mosleh and Apostolakis, 1985) as it is rigorously derived from explicit assumptions which clarify the role of conditional independence. It is similar to the solution proposed in Hora and Iman (1990), although the latter does not use contamination parameters. The main conclusions of this review are:

- the assumptions underlying the model are sensible, if not inevitable;

- the derivations in Pörn (1990) are correct;

- improper hyperpriors are problematic as expectations are infinite, even after updating on observations. The statistical consistency of the model is sensitive to the manner in which these improper hyperpriors are truncated, and the heuristic suggested by Pörn would tend to render the model inconsistent.

This section is written for the specialist interested in mathematical aspects of the model, and some issues are treated here in more detail than in Pörn (1990).

### 2.8.1   Mathematical background

The Poisson distribution with rate $\lambda$ and time duration $T$ is concentrated on the non-negative integers, and the probability associated with integer $n$ is

$$e^{-\lambda T}(\lambda T)^n/n! \qquad \lambda > 0, \quad T > 0$$

Its expectation and variance are $\lambda T$.

The gamma distribution $G(\lambda \mid \alpha, \beta)$ in $\lambda$ with shape $\alpha$ and scale $\beta$ has the density

$$\lambda^{\alpha-1} e^{-\beta\lambda} \beta^\alpha / \Gamma(\alpha); \alpha > 0, \beta > 0; \Gamma(\alpha) = \int_0^\infty u^{\alpha-1} e^{-u} du$$

$\Gamma(\alpha)$ is the gamma function, for $n$ an integer, $\Gamma(n) = (n-1)!$. Where E denotes expectation, Var denotes variance, SD denotes standard deviation,

CF denotes coefficient of variation, and Mode is the highest point of the density function:

$$E(G(\lambda \mid \alpha, \beta)) = \alpha/\beta$$

$$\mathrm{Var}(G(\lambda \mid \alpha, \beta)) = \alpha/\beta^2$$

$$\mathrm{SD}(G(\lambda \mid \alpha, \beta)) = \sqrt{\alpha}/\beta$$

$$\mathrm{CF}(G(\lambda \mid \alpha, \beta)) = \mathrm{SD}/\mathrm{E} = 1/\sqrt{\alpha}$$

$$\mathrm{Mode}(G(\lambda \mid \alpha, \beta)) = (\alpha - 1)/\beta$$

The beta density on the interval $[0, 1]$ with parameters $a > 0$, $b > 0$ is given by

$$u^{a-1}(1 - u)^{b-1}/B(a, b)$$

where

$$B(a, b) = \Gamma(a)\Gamma(b)/\Gamma(\alpha + \beta)$$

Its expectation and variance are, respectively, $a/(a + b)$ and $ab/[(a + b + 1)(a + b)^2]$. We recall that random variables $X$ and $Y$ are independent conditional on random variable $Z$ if and only if

$$P(XY \mid Z) = P(X \mid Z)P(Y \mid Z)$$

or, equivalently,

$$P(X \mid Y, Z) = P(X \mid Z)$$

If $X$ and $Y$ are conditionally independent given both $Z_1$ and $Z_2$, i.e. given $\{Z_1, Z_2\}$, it does *not* follow that $X$ and $Y$ are conditionally independent given $Z_1$. Similarly, if $X$ and $Y$ are conditionally independent given $Z_1$, and are conditionally independent given $Z_2$, it does *not* follow that they are conditionally independent given $\{Z_1, Z_2\}$.

## 2.8.2 Assumptions

We consider a population of similar plants, where operational data for a given component type is generated by each plant. Each plant may have several component sockets for the given component type. The assumptions given below are not taken verbatim from Pörn (1990) but capture the intent.[3] We introduce the following notation:

$$\underline{x}_n = (x_1, \ldots x_n): \text{ numbers of failures at plants } 1, \ldots, n$$

$$\underline{T}_n = (T_1, \ldots T_n): \text{ operating times at plants } 1, \ldots, n$$

---

[3] Pörn does not discuss component sockets and does not suggest that sockets from the same plant are independent and identical realizations of the same Poisson process. Nor does he discuss competing causes of service outage. However, these assumptions underlie the use of the model in analysing the T-Book data.

$(\lambda_1, ...\lambda_n)$: realizations of random variables $(\Lambda_1, ...\Lambda_n)$ at plants $1, .., n)$

$\theta = (\alpha, \beta, c)$: (hyper)parameters of the distribution of $\Lambda$

1. Each plant–component type is associated with a realization $\lambda$ of a random variable $\Lambda$. At a given plant and regardless of information from other plants (i.e. given $\lambda$ and given any information from other plants), the failure times from each component socket for the given component type are independent, and follow a Poisson distribution with parameter $\lambda$; moreover, the failure process is independent of other causes of service outage.

   To conform with the notation in Pörn (1990), we use the symbol $\lambda_i$ to denote both the random variable $\Lambda_i$ associated with plant $i$, and the realization of this random variable at plant $i$. The parameters $\theta$ are unknown, and uncertainty over possible values of $\theta$ will be described by a subjective distribution. Mathematically this simply means that $\theta$ is a random vector.

2. $(\lambda_1, ...\lambda_n)$ are conditionally independent and identically distributed given $\theta$.

3. $(x_1, ... x_n)$ are conditionally independent given $(\theta, \lambda_1, ...\lambda_n)$.

4. Given $\lambda_i$, $x_i$ and $(\theta, \lambda_1, ...\lambda_{i-1}, \lambda_{i+1}, ...\lambda_n)$ are conditionally independent.

5. $P(\lambda_i | \theta) = G(\lambda_i | \alpha, \beta)(1 - c) + G(\lambda_i | \frac{1}{2}, 1)c$, where $c$ is a random variable taking values in $[0, 1]$.

Assumption 5 describes a 'contaminated gamma' prior. $\lambda_i$ follows a gamma distribution with shape and scale of $\alpha$ and $\beta$, respectively, 'contaminated' with a relatively vague gamma with random mixing coefficient $c$.

### 2.8.3   The hyperpriors

A hyperprior distribution must be placed on $\theta = (\alpha, \beta, c)$. First, $\theta$ is re-parametrized into variables which are assumed to be independent. For the uncontaminated term $G(\lambda_i | \alpha, \beta)$, the coefficient of variation $v = 1/\sqrt{\alpha}$ is assumed to be independent of the expected number of failures at time $T$, given $\alpha$ and $\beta$; $\mu' = T\alpha/\beta$. Both are believed to be independent of $c$.

The parameters $v$, $\mu'$ and $c$ are assigned non-informative hyperprior distributions. Non-informative priors (Box and Tiao, 1973) are simply the uniform distribution with respect to a transformation of the parameter under which the likelihood is 'data translated'. If $r_v$ is a function of the data which estimates $v$, then the likelihood function $L(r_v, v, \mu', c)$ is data translated for $v$ if

$$\partial L(r_v, v, \mu', c)/\partial v = \partial L(r_v, v, \mu', c)/\partial r_v$$

It is obvious that the property of being data translated for $v$ depends, in general, on the values of the other parameters $\mu'$ and $c$. Hence it is not really possible for the non-informative hyperpriors to be independent; rather, independence must be regarded as an assumption made for convenience. This, of course, is quite sensible so long as the effects of the assumption are small. Pörn (1990) provides arguments for choosing the following non-informative (improper) densities for the parameters $v$, $\mu'$ and $c$:

$$g(v) = 1/v \qquad v > 0$$

$$k(\mu') = 1/(\mu'(1 + \mu'))^{\frac{1}{2}} \qquad \mu' > 0$$

$$j(c) = 1/c^{\frac{1}{2}} \qquad c \in [0, 1]$$

As is often the case with non-informative priors, the densities for $v$ and $\mu'$ are improper, that is, their integral is infinite. Proper densities are obtained by truncating the range of the parameters. We note that $g$ is the improper loguniform density, and $j$ is the proper density of a squared uniform variable on $[0, 1]$.

A joint hyperprior density for $\alpha$, $\beta$, $c$ is obtained by transforming back to the hyperparameters $\alpha$ and $\beta$:

$$\alpha = 1/v^2 \qquad \beta = T/(v^2\mu')$$

The Jacobian of the transformation is

$$J = \mathrm{Det}^{-1}\begin{bmatrix} \partial\alpha/\partial v & 0 \\ \partial\beta/\partial v & \partial\beta/\partial\mu \end{bmatrix} = \frac{v^5\mu'^2}{2T}$$

Transforming $v, \mu'$ to $\alpha, \beta$, the independent improper density $g(v)k(\mu')$ is transformed into

$$g(v(\alpha))k(\mu'(\alpha, \beta))|J| = \frac{1}{\beta(\alpha(\alpha + \beta)/T)^{1/2}}$$

The transformed density is also improper.

## 2.8.4   Discussion

The most important assumption is undoubtedly the first. It implies:

1. the outages due to preventive maintenance, degraded or incipient failure and the like contain no information about the failure rate;

2. the failure rates $\lambda_i$ do not change with time;

3. the data from different plants for the same component type cannot be pooled, while the data from different component sockets can be pooled;

4. the only information from each plant which need be retained and recorded is the number of failures and the total time in operation.

The last point becomes clear in the next section where the inference model is discussed.

The distribution of $\Lambda$ is characteristic of the *population* of which the component type is considered to be a member. If this population changes, the distribution of $\Lambda$ changes and the assessment of $\lambda_i$ changes. This is quite familiar when probability is interpreted as limiting relative frequency with respect to a reference class; changing the reference class changes the probability. For example the 'probability' that a male lives longer than 65 years is different from the 'probability' that a smoker lives longer than 65 years. For the component type 'inner isolation valve', we may consider the population to be the inner isolation valves at all reactors, or we may consider the population to be the set of valves (inner, outer, pilot, check) at a given reactor. The model gives no guidance how to choose the population, but only tells how to process data once a population is chosen. The assessment of the failure rate for the inner isolation valve will not be the same when referred to different populations. Although this may seem paradoxical at first, it is a natural consequence of the decision to learn about a component type from a population of which it is a member.

Within the frequentist school, there has been an extended discussion (Hempel, 1968) about how to choose a reference class for given purposes. A similar discussion would arise if Pörn's model were aggresively challenged on this point. Why should the failure rate for inner isolation valves at Ringhals be updated with data from inner isolation valves at Baerseback but not with data from outer isolation valves at Ringhals? What principles guide these choices?

The distribution of $\theta$ describes the uncertainty in $\lambda_i$ prior to any observations. If this distribution can be chosen on *a priori* grounds, then nothing more need be said. In the following section, we show that for some choice of tractable improper hyperpriors (the hyperpriors given by Pörn are not tractable for this purpose):

1. the model is not well-behaved as the posterior expectation does not exist, even after arbitrarily many observations;

2. the statistical consistency of the model is sensitive to the method of truncating the improper hyperpriors;

3. the heuristic interpretation of the hyperparameters encourages a method of truncation which defeats statistical consistency.

Further, we do not find the motivation for introducing the contamination distribution $G(\lambda_i | \frac{1}{2}, 1)$ overwhelming. Indeed, if the arguments for introducing non-informative hyperpriors on $\alpha$ and $\beta$ carry any weight, then the same arguments would unmotivate the introduction of $G(\lambda_i | \frac{1}{2}, 1)$; indeed, the parameters of this distribution can be seen as very informative choices for $\alpha$ and $\beta$.

The assumptions regarding conditional independence appear reasonable. We regard the identification of these assumptions and the derivation of an inference model as a significant achievement.

## 2.9    The inference model

The problem is to determine an uncertainty distribution for a component type failure rate at a given plant, using information from other plants. Let the plant of interest be the $(n+1)$th plant, and suppose that $T_1 = T_2 = \ldots = T_{n+1}$. The problem may then be described as follows: determine $P(\lambda_{n+1}|\underline{x}_{n+1})$. The argument involves a subtle use of the conditional independence assumptions, and we go through the steps in detail, though omit the proofs (see Cooke *et al.*, 1995).

*Lemma 1*    Letting $\propto$ denote proportionality:

$$\text{(i)}\quad P(\theta|\underline{x}_n) \propto \prod_{i=1}^{n}\left[\int P(x_i|\lambda_i)P(\lambda_i|\theta)\,\mathrm{d}\lambda_i\right]P(\theta)$$

$$\text{(ii)}\quad P(\lambda_{n+1}|\underline{x}_n, \theta) = P(\lambda_{n+1}|\theta)$$

Lemma 1(i) entails that the $x_i$ are conditionally independent given (only) $\theta$, since (using assumption 4)

$$P(\theta|\underline{x}_n) \propto \left[\prod_{i=1}^{n}\int P(x_i|\lambda_i)P(\lambda_i|\theta)\,\mathrm{d}\lambda_i\right]P(\theta)$$

$$= \left[\prod_{i=1}^{n}\int P(x_i|\theta, \lambda_i)P(\lambda_i|\theta)\,\mathrm{d}\lambda_i\right]P(\theta) = \prod_{i=1}^{n}P(x_i|\theta)P(\theta)$$

Lemma 1 says that $x_1$ can influence our beliefs about $x_2$ *only* by influencing our beliefs about $\theta$. If $\theta$ is assumed to be known, then there is nothing which data from one plant can tell us about the failure rate at another plant.
    Continuing, we have

$$P(\lambda_{n+1}|\underline{x}_{n+1}) = \int P(\lambda_{n+1}|\underline{x}_{n+1}, \theta)P(\theta|\underline{x}_{n+1})\,\mathrm{d}\theta$$

Applying Bayes' theorem:

$$P(\lambda_{n+1}|\underline{x}_{n+1}, \theta) = P(x_{n+1}|\underline{x}_n, \lambda_{n+1}, \theta)P(\lambda_{n+1}|\underline{x}_n, \theta)/P(x_{n+1}|\underline{x}_n, \theta)$$

and

$$P(\theta|\underline{x}_{n+1}) = P(x_{n+1}|\underline{x}_n, \theta)P(\theta|\underline{x}_n)/P(x_{n+1}|\underline{x}_n)$$

Substituting these expressions into the first equation yields

$$P(\lambda_{n+1}|\underline{x}_{n+1}) = \int P(x_{n+1}|\underline{x}_n, \lambda_{n+1}, \theta)P(\lambda_{n+1}|\underline{x}_n, \theta)P(\theta|\underline{x}_n)\,\mathrm{d}\theta/P(x_{n+1}|\underline{x}_n)$$

To reduce the first term under the integral, note that

$$P(x_{n+1}|\underline{x}_n, \lambda_{n+1}, \theta) = \int P(x_{n+1}|\underline{x}_n, \lambda_1 \ldots \lambda_{n+1}, \theta)P(\lambda_1 \ldots \lambda_n|\lambda_{n+1}, \underline{x}_n, \theta)$$
$$\mathrm{d}\lambda_1 \ldots \mathrm{d}\lambda_n$$

By assumptions 3 and 4,

$$P(x_{n+1}|\underline{x}_n, \lambda_1 \dots \lambda_{n+1}, \theta) = P(x_{n+1}|\lambda_1 \dots \lambda_{n+1}, \theta) = P(x_{n+1}|\lambda_{n+1})$$

We may now integrate out over $\lambda_1, \dots, \lambda_n$ to obtain

$$P(x_{n+1}|\underline{x}_n, \lambda_{n+1}, \theta) = P(x_{n+1}|\lambda_{n+1})$$

By lemma 1(ii),

$$P(\lambda_{n+1}|\underline{x}_n, \theta) = P(\lambda_{n+1}|\theta)$$

Together, these yield

$$P(\lambda_{n+1}|\underline{x}_{n+1}) = P(x_{n+1}|\lambda_{n+1}) \int P(\lambda_{n+1}|\theta) P(\theta|\underline{x}_n) \, d\theta / P(x_{n+1}|\underline{x}_n)$$

The term $P(x_{n+1}|\underline{x}_n)$ depends only on the observed data and may be absorbed into a constant of proportionality. To reduce this further, we apply lemma 1(i). The integrals inside the brackets can be evaluated with assumption 5. We find, after some calculation,

$$\int P(x_i|\lambda_i) P(\lambda_i|\theta) d\lambda_i = \frac{\Gamma(x_i + \alpha)}{\Gamma(x_i + 1) \Gamma(\alpha)} \left(\frac{\beta}{\beta + T_i}\right)^\alpha \left(\frac{T_i}{\beta + T_i}\right)^{x_i} (1 - c)$$

$$+ \frac{\Gamma(x_i + 1/2)}{\Gamma(x_i + 1) \Gamma(1/2)} \left(\frac{1}{1 + T_i}\right)^{1/2} \left(\frac{T_i}{1 + T_i}\right)^{x_i} c$$

With assumption 1 we have

$$P(x_{n+1}|\lambda_{n+1}) = \exp\{-\lambda_{n+1} T_{n+1}\} (\lambda_{n+1} T_{n+1})^{x_{n+1}} / \Gamma(x_{n+1} + 1)$$

We term the above expression the 'likelihood of $\lambda$'. Combining this with assumption 5, we have

$$P(\lambda_{n+1}|\underline{x}_{n+1}) = G(\lambda_{n+1}|\alpha + x_{n+1}, \beta + T_{n+1})(1 - c)$$

$$+ G(\lambda_{n+1}|\tfrac{1}{2} + x_{n+1}, 1 + T_{n+1})c$$

where $\theta = (\alpha, \beta, c)$ follows a distribution which is proportional to

$$\prod_{i=1}^{n} \left[ \frac{\Gamma(x_i + \alpha)}{\Gamma(x_i + 1) \Gamma(\alpha)} \left(\frac{\beta}{\beta + T_i}\right)^\alpha \left(\frac{T_i}{\beta + T_i}\right)^{x_i} (1 - c) \right.$$

$$\left. + \frac{\Gamma(x_i + 1/2)}{\Gamma(x_i + 1) \Gamma(1/2)} \left(\frac{1}{1 + T_i}\right)^{1/2} \left(\frac{T_i}{1 + T_i}\right)^{x_i} c \right] P(\alpha, \beta, c)$$

We refer to the above expression as the 'likelihood in $\alpha, \beta, c$'. The entire inference model can be expressed in prose as

Posterior$(\lambda|\underline{x}) \propto$ likelihood$(\lambda|x_{n+1}, \alpha, \beta)$

$$\times \text{likelihood } (\alpha, \beta, c|\underline{x}_n) \times \text{prior}(\alpha, \beta, c)$$

### 2.9.1   Simple Bayesian model

For the purposes of comparison, it is useful to consider the simple Bayesian Poisson–gamma model. On this model each plant is treated separately and gives no information about other plants. The failure rate $\lambda_i$ at plant $i$ follows a gamma distribution $G(\lambda \mid \alpha, \beta)$. In the simple Bayesian model, we have

$$\text{posterior}(\lambda \mid \text{data}) \propto \text{likelihood}(\lambda \mid \text{data}) \times \text{prior}(\lambda)$$

After observing $x$ failures in operation time $T$, the posterior or updated distribution for $\lambda$ is

$$P(\lambda \mid x, T) \propto P(x, T \mid \lambda)P(\lambda) \propto e^{-\lambda T}(\lambda T)^x \lambda^{\alpha-1} e^{-\beta\lambda} \propto G(\lambda \mid \alpha+x, \beta+T)$$

Had we started from the improper prior $G(\lambda \mid 0,0) \sim 1/\lambda$, the updated distribution would be simply $G(\lambda \mid x, T)$. For this reason, the parameters $\alpha$ and $\beta$ are often interpreted in terms of 'virtual observations'; adopting the prior $G(\lambda \mid \alpha, \beta)$ is equivalent to having started with $G(\lambda \mid 0,0)$ and having performed a 'virtual observation' of $\alpha$ failures in $\beta$ operational time. Note that the improper prior becomes a proper prior after updating with one observed failure. Whereas the expectation of $G(\lambda \mid 0,0)$ does not exist, the expectation of $G(\lambda \mid 1, T)$ does exist. This fact justifies the use of improper priors in simple cases like these.

The same result may be obtained by another argument. Suppose the number of failures $X$ in time $T$ follows a Poisson distribution with parameter $\lambda$. We have $E(X) = Var(X) = \lambda T$ and $E(X/T) = \lambda$, $Var(X/T) = \lambda/T$. If we now choose a gamma prior satisfying

$$E(G(\lambda \mid \alpha, \beta)) = \alpha/\beta = E(X/T) = \lambda$$

$$Var(G(\lambda \mid \alpha, \beta)) = \alpha/\beta^2 = Var(X/T) = \lambda/T$$

Then, upon substituting the maximum likelihood estimate $x/T$ for $\lambda$, we find that $\alpha = x$, $\beta = T$. The result of this simple moment fitting is the same as updating the improper prior. For 'reasonable' values of $x$ and $T$, the 5 per cent and 95 per cent quantiles of $G(\lambda \mid x, T)$ are almost identical to the lower and upper 5 per cent classical confidence bounds.

### 2.9.2   Asymptotic properties

It is important to understand the behaviour of the inference model as the observation times (and also the numbers of failures) get large. By assumption, as $T_i \rightarrow \infty$, we have $x_i/T_i \rightarrow \lambda_i$, at each plant $i$. The inference model is consistent in a statistical sense if [4]

$$P(\lambda_{n+1} \mid \underline{x}_{n+1} \rightarrow \delta_{\lambda_{n+1}} \text{ as } T_{n+1}$$

---

[4] The type of convergence meant here is weak convergence, i.e. convergence of the cumulative distribution functions at points of continuity. A sufficient condition for consistency is $E(\lambda_i \mid \underline{x}) \rightarrow \lambda_i$, $Var(\lambda_i \mid \underline{x}) \rightarrow 0$.

where $\delta_{\lambda_{n+1}}$ denotes the Dirac measure concentrated at $\lambda_{n+1} = \lim_{T_{n+1} \to \infty} (x_{n+1}/T_{n+1})$. Consistency entails that $E(\lambda_{n+1} | \underline{x}_{n+1}) \to \lambda_{n+1}$ as $T_{n+1} \to \infty$. Note that this statement says nothing about $T_j, j \neq n + 1; T_j$ might increase at a different rate from $T_{n+1}$.

We note first that if $\alpha$ and $\beta$ are concentrated on a bounded (non-negative) set and $x_{n+1}$ and $T_{n+1}$ are sufficiently large, then

$$P(\lambda_{n+1} | \underline{x}_{n+1}) \sim G(\lambda_{n+1} | x_{n+1}, T_{n+1})$$

where $\sim$ denotes '...converge weakly to the same distribution as...'. Since Var $G(\lambda_{n+1} | x_{n+1}, T_{n+1}) \to 0, EG(\lambda_{n+1} | x_{n+1}, T_{n+1}) = x_{n+1}/T_{n+1} \to \lambda_{n+1}$ as $T_{n+1} \to \infty$; the model is clearly consistent in this case, although the restriction on the support of $\alpha$ and $\beta$ is essential.

Equally obvious is the following remark: if $\beta = T$ and $\alpha \neq x_{n+1}$ then, barring a miraculous good fortune with the contamination distribution, $E(\lambda_{n+1} | \underline{x}_{n+1}) \neq x_{n+1}/T$. Hence, if the hyperprior for $\alpha$ and $\beta$ is chosen *after* observing $x_i$ and $T$, and is chosen such that $\alpha \sim x_i, \beta \sim T$, then this renders the model inconsistent, unless $x_i = x_{n+1}$.

The following section contains a technical analysis of improper hyperpriors and the result of updating the hyperpriors. It provides insight into the working of the inference model, but may be skipped without loss of continuity.

### 2.9.2.1   Hyperpriors and hyperposteriors

With the simple Bayesian model, improper priors are justified by th e fact that the improper priors become proper after updating on one failure. The asymptotic behaviour of the model is not influenced by the improper prior. We are interested as to whether similar features hold for the inference model of Section 2.9.

We study the 'uncontaminated inference model' by putting $c = 0$. Assuming that $n = 1$ and recalling that $\Gamma(k) = (k - 1)!$ for any positive $k$ (this defines the factorial function for non-integer $k$), and dropping the index $i$, we may write the likelihood in $\alpha, \beta$ as

$$L(\alpha, \beta | x, T) = \frac{\Gamma(x + \alpha)}{\Gamma(x + 1) \Gamma(\alpha)} \left( \frac{\beta}{\beta + T} \right)^\alpha \left( \frac{T}{\beta + T} \right)^x$$

$$= \binom{x + \alpha - 1}{x} \left( \frac{\beta}{\beta + T} \right)^\alpha \left( \frac{T}{\beta + T} \right)^x$$

where

$$\binom{x + \alpha - 1}{x} = \frac{(x + \alpha - 1)!}{x! \, (\alpha - 1)!}$$

The asymptotic behaviour of the 'hyperposterior' $P(\alpha, \beta | x, T) P(\alpha, \beta)$ will essentially be determined by the maximum of $L(\alpha, \beta | x, T)$. The significant fact is that $L(\alpha, \beta \,|\, x, T)$ has no maximum; it is asymptotically maximal along a ridge. This explains the 'persistence' of improper hyperpriors to be studied below. Setting the derivative with respect to $\beta$ equal to zero we find $\alpha/\beta = x/T$. We use this to eliminate $\beta$ from the likelihood in $\alpha, \beta$. Writing $\alpha^* = \alpha/x$, this likelihood becomes

$$L(\alpha^*, T\alpha^* \,|\, x, T) = \binom{x + x\alpha^* - 1}{x} \left(\frac{\alpha^*}{1 + \alpha^*}\right)^{x\alpha^*} \left(\frac{1}{1 + \alpha^*}\right)^{x}$$

The derivative of this function with respect to $\alpha^*$ is a rather complicated expression, but is positive for $\alpha^* > 0$, $x > 0$, as can be checked with MAPLE (see Fig. 2.7). This means that $L(\alpha, \beta \,|\, x, T)$ increases along the ridge $\beta = T\alpha^*$. The limiting value is given by:

**Lemma 2:**   $\lim\limits_{\alpha^* \to \infty} \binom{x + x\alpha^* - 1}{x} \left(\dfrac{\alpha^*}{1 + \alpha^*}\right)^{x\alpha^*} \left(\dfrac{1}{1 + \alpha^*}\right)^{x} = (2\pi x)^{-1/2}$

We first explore the behaviour of the posterior in $\lambda_2$, $P(\lambda_2 \,|\, \underline{x}_2)$, when $\alpha$ and $\beta$ are assigned uniform improper hyperpriors ($\alpha$ need not be restricted to integers), and when $T_1 = T_2$. Consistent with assumption 1, we assume that $x_1/T \to \lambda_1$, $x_2/T \to \lambda_2$, as $T \to \infty$. Conditional on $\alpha$, and using the uniform hyperprior $\mathrm{d}P(\beta) = \mathrm{d}\beta$, the 'hyperposterior' density in $\beta$ takes the form

$$P(\beta \,|\, \alpha, x_1, T)\, \mathrm{d}\beta \propto \left(\frac{\beta}{\beta + T}\right)^{\alpha} \left(\frac{T}{\beta + T}\right)^{x_1} \mathrm{d}\beta$$

Under the transformation $u = T/(T + \beta)$, $\beta = T(1 - u)/u$, $\mathrm{d}\beta = -T\mathrm{d}u/u^2$, we may write:

$$P(u \,|\, \alpha, x, T)\mathrm{d}u \propto (1 - u)^{\alpha}(u)^{x_1 - 2}\mathrm{d}u$$

$$P(\lambda_2 \,|\, \underline{x}_2) = \int_0^{\infty} \int_0^1 G(\lambda_2 \,|\, \alpha + x_2, T/u)\, P(u \,|\, \alpha, x, T)\, \mathrm{d}u\, \mathrm{d}\alpha$$

$P(u \,|\, \alpha, x, T)$ is a Beta density with mean $(x_1 - 1)/(\alpha + x_1)$. If this model were well behaved, then we should be able to compute the posterior expectation of $\lambda_2$, and interchange the order of integration of $\lambda_2$, $\alpha$ and $u$. The expectation of $G(\lambda_2 \,|\, \alpha + x_2, T/u)$ is $(x_2 + \alpha)u/T$. Integrating this with respect to u:

$$\int_0^{\infty} \int_0^1 \frac{(x_2 + \alpha)\, u\, (1 - u)^{\alpha}(u)^{x_1 - 2}}{TB(\alpha + 1, x_1 - 1)}\, \mathrm{d}u\, \mathrm{d}\alpha$$

$$= \int_0^{\infty} (x_2 + \alpha)\, B(\alpha + 1, x_1)/[TB(\alpha + 1, x_1 - 1)]\, \mathrm{d}\alpha$$

$$= \int_0^{\infty} (x_2 + \alpha)(x_1 - 1)/[T(x_1 + \alpha)]\, \mathrm{d}\alpha$$

The latter integral is not finite. Indeed,

$$(x_2 + \alpha)(x_1 - 1)/[T(x_1 + \alpha)] \to (x_1 - 1)/T = \lambda_1 - 1/T \qquad \text{as } \alpha \to \infty$$

This result yields the following interpretation. The inference model is not well behaved for uniform hyperpriors on $\alpha$ and $\beta$, as the posterior expectation of $\lambda_2$ does not exist, even after observing $(x_1, T)$. This is basically because the hyperposterior does not 'peak' but 'ridges'. Obviously, the model is not consistent in this case. Truncating the hyperpriors, at least $\alpha$, ensures finite expectations. When the method of truncation may depend on $x_i$ and $T$, the asymptotic behaviour of the model depends essentially on how the improper hyperpriors are truncated. We distinguish a number of possibilities for $\alpha$ ($\beta$ has always the uniform hyperprior):

1. If $T$ is very large, and if $\alpha$ is truncated to very large values, then $E(\lambda_2 | \underline{x}_2) \approx \lambda_1$.

2. If $\alpha$ is small relative to $x_1$ (and hence also $T$) then $E(\lambda_2 | \underline{x}_2) \approx \lambda_2$.

3. If $\alpha = x_1$ then $E(\lambda_2 | \underline{x}_2) \approx (x_1 + x_2)/2T = EG(\lambda_2 | x_{\text{average}}, T) \sim (\lambda_1 + \lambda_2)/2$.

4. If $\alpha$ has a proper distribution with density $f$ (perhaps depending on $x_i$), then $E(\lambda_2 | \underline{x}_2) = \int [(x_1 - 1)/T][(x_2 + \alpha)/(x_1 + \alpha)]f(\alpha)d\alpha$.

The model is consistent only in case 2. All of these results assume that a uniform improper hyperprior is assigned to $\beta$.

### 2.9.3   Method of truncation

Given the importance attaching to the method of truncation, it is most desirable to motivate the method employed and perform sensitivity analysis. Supplying such motivation is not one of the strong points of the Bayesian approach. The best we can do is to provide a heuristic interpretation of the hyperparameters and then appeal to common sense. Pörn suggests a heuristic for $\mu'$:

$$\mu': \text{the number of expected failures in time } T = \sum_{i=1}^{n} T_i/n$$

No heuristic is indicated for $v$, beyond saying that it is the coefficient of variation of our uncertainty in $\lambda_i$ prior to observation. Now, the heuristic for $\mu'$ suggests that $\mu'$ is of the same order as the $x_i$, which is of order $\lambda_i T$. The coefficient of variation is the standard deviation of $\lambda_i$ divided by the expectation of $\lambda_i$. Estimating these with simple moment fitting from data $x_i$, $T$, as discussed in Section 2.9.1, would suggest that the expectation of $\lambda_i$ is of order $\mu'/T$ and the variance of $\lambda_i$ is of order $\lambda_i/T \sim \mu'/T^2$. Hence the order of $v$, $\alpha$ and $\beta$ is given by

$$v \sim 1/\sqrt{\mu'} \sim 1/\sqrt{x_i}$$

$$\alpha = 1/v^2 \sim x_i$$

$$\beta = \alpha T/\mu' \sim x_i T/(\lambda_i T) \sim T$$

If we apply Pörn's heuristic using $\mu' \sim$ numbers of failures in $T^*$, and think of $v$ in terms of moment fitting as in Section 2.9.1, then we are led to truncate the hyperpriors in a way which threatens the statistical consistency of the model. The order of $\alpha$ and $\beta$ is that of case 3 in Section 2.9.2.1, and the influence of $\alpha$ and $\beta$ does not die off as the observation time increases.

Note also that the order relation $\nu \sim 1/\sqrt{\mu'}$ would also suggest a strong negative correlation between the uncertainty distributions over $\nu$ and $\mu'$.

The simple Bayesian model of Section 2.9.1 suggests that $\alpha$ be interpreted as a virtual number of failures in virtual observation time $\beta$, where these virtual numbers are chosen to reflect our expected value for $\lambda$, $\alpha/\beta$, and our variance about this expected value, $\alpha/\beta^2$. Under this interpretation, it is unreasonable that the observed $x_1$ and $T$ determine the ranges of $\alpha$ and $\beta$.

Taking account of the contaminating random variable $c$ would vastly complicate the computations, but would not affect the conclusion. The model would still be inconsistent for uniform improper priors, and the range on which $\alpha$ were truncated would essentially determine the posterior expectation of $\lambda_2$.

### 2.9.4   Numerical example

Let $x_1 = 10$, $T = 1000$. The computations and figures below are made with MAPLE which does not support Greek letters, so $\alpha$ and $\beta$ are denoted $a$ and $b$. Fig. 2.5 shows the maximum of $L(a, \beta \mid x_1, T)$ as a function of $a$. Note

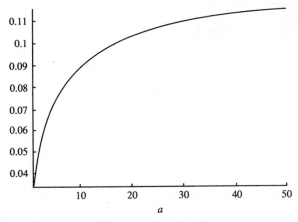

**Fig. 2.5** Max $L\,(a, b \mid x_1, T)$ as a function of $a$

that the maximum converges to the limiting value $(2\pi x_1)^{-\frac{1}{2}} = 0.125$. Fig. 2.6 shows the likelihood $L(a, \beta \mid x_1, T)$ for $\alpha = 1, .., 50$, $\beta = 100, .., 5000$. Fig. 2.7 shows $L(a, \beta \mid x_1, T)$ for $x_1 = 100$, $T = 10\,000$, and $\alpha = 1, .., 50$, $\beta = 100, .., 5000$. Note that the likelihood has become more concentrated about $\alpha/\beta = x_1/T$. The non-informative prior for $\alpha, \beta$ is given in Fig. 2.8. Fig. 2.9 shows the hyperposterior, 'likelihood × prior' with $x_1 = 10$. The 'ridging' of the likelihood $L(a, \beta \mid x_1, T)$ is not very pronounced.

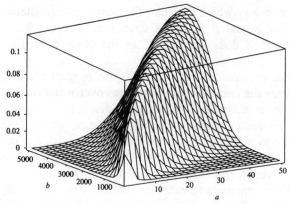

**Fig. 2.6** $L\,(a, b \mid x_1, T)$; $a = 1...50$, $b = 100...5000$, $x_1 = 10$, $T = 1000$

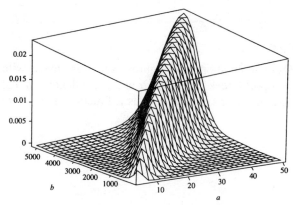

**Fig. 2.7** $L\,(a, b \mid x_1, T)$; $a = 1...50$, $b = 100...5000$, $x_1 = 100$, $T = 10\,000$

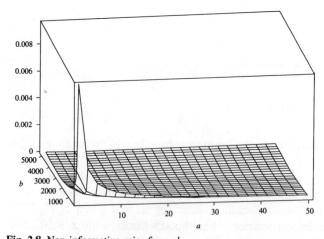

**Fig. 2.8** Non-informative prior for $a$, $b$

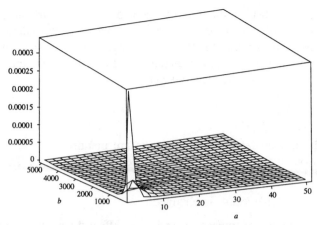

**Fig. 2.9** Hyperposterior for $a$, $b$; $x_1 = 10$, $T = 1000$

### 2.9.5   Verification tasks

On the basis of the above discussion, we identify the calculations with the Pörn model which will contribute to verification.

- Task 1 – determine choices of hyperparameter ranges yielding results in agreement with those reported in Pörn (1990).

- Task 2 – compare results of task 1 with classical and simple Bayesian methods.

- Task 3 – compare results of task 1 with results when ranges of hyper-parameters are shifted.

- Task 4 – compare the results of task 1 with the 'uncontaminated model'.

- Task 5 – compare results of task 1 when dependence is assumed in the hyperprior.

- Task 6 – compare results of task 1 when data from different plants are pooled prior to processing.

## 2.10   Coding the inference model

To perform the verification tasks identified in Section 2.9, the inference model has been coded in two independent ways. The first method involves writing the formulae of the inference model in the higher-order programming language MAPLE. The integrals over $\alpha$, $\beta$ and $c$ are evaluated numerically to determine $P(\lambda_{n+1} | \underline{x}_{n+1})$ for a large number of values of $\lambda_{n+1}$. The resulting numbers are normalized to approximate the posterior distribution of $\lambda_{n+1}$. Running the code for 100 values of $\lambda_{n+1}$ with a high degree of numerical accuracy may take several hours, but rough results may be obtained in 15 min on a fast PC. This is adequate for most of the verification tasks identified above.

To explore the effects of dependence in the hyperpriors, the model is evaluated with the program UNICORN developed at the TU Delft.

UNICORN simulates the distribution of functions of random variables, where the distributions of the random variables may be dependent. UNICORN can be used to perform Monte Carlo integration in which the marginal prior distributions are rank-correlated and fitted to a maximal entropy joint distribution. For details see Cooke *et al.* (1995).

## 2.11    Results

This section reproduces calculations whose results are mentioned in Pörn (1990), and carries out the verification tasks defined in Section 2.9. The main conclusions of this are as follows:

- good agreement with Pörn's results is obtained;

- different defensible choices of truncation ranges can affect the median and 95 per cent quantile by a factor 5;

- the contamination feature is not important for the data analysed here;

- negative correlation between $v$ and $\mu'$ can increase the posterior median and 95 per cent quantile by a factor 1.5;

- the decision how to pool 'other plants' is less important than the decision what to consider as the 'current plant', i. e. the pooling of sockets at a given plant. Increasing aggregation of other plants leads to results similar to the simple Bayes model; including the current plant approaches the simple Bayes model for the entire population.

The example used by Pörn (1990) to illustrate the inference model uses data for centrifugal pumps with flow rate 75–150 kg/s reproduced in Table 2.2. Classical maximum likelihood estimates and 90 per cent classical confidence bounds are also shown, when computable. This data has been chosen as test data for performing the verification tasks.

The $(n+1)$th component, $x_{15}$, is assumed to have run for 5000 h and experienced three failures. The simple maximum likelihood estimate for $\lambda_{15}$ would be $x_{15}/T = 0.0006$. The inference model of Section 2.9 uses information from the other 14 plants. Pörn reports the results in Table 2.3 for the posterior distribution $P(\lambda_{15}|\underline{x}_{n+1})$. Task 1 (Section 2.9.5) is to determine choices of hyperparameter ranges yielding results in agreement with those reported in Pörn (1990). Table 2.4 shows the result of the MAPLE calculation when $\alpha$ and $\beta$ are restricted as suggested by Pörn (1990, p.26). The agreement with Pörn's results is good, considering that Pörn's exact choice of parameters and numerical integration technique is not given in Pörn (1990). The computation with $8 \times 8 \times 8$ gridpoints took 3 h on a fast PC (anno 1994); the computation with $4 \times 4 \times 4$ gridpoints took 0.5 h. The results with UNICORN are also shown. Given the different range of $\beta$, the approximation of the gamma distribution, the number of samples (2300) and the very different method employed by UNICORN, the agreement is about as good as can be expected. The distribution over $\beta$ was obtained by truncating $\mu'$ to $[0, 15]$. The computation took about 6 h; the computation with 134 samples took 0.5 h.

Task 2 (Section 2.9.5) is to compare the results of task 1 with classical and simple Bayesian methods. In the classical and simple Bayes analysis, we may *either* use the data from plant 15 in isolation *or* pool the data from all 15 plants, as if all data were generated by the same failure rate. This data would lead to the estimates of $\lambda_{15}$ shown in Table 2.5. We see that the results of the classical and the simple Bayes analysis for both the pooled and unpooled data are almost identical, as indicated in Section 2.9.1. Relative to Table 2.3, the main effect is to raise the median and 95 per cent quantiles by roughly a factor of 2.

**Table 2.2** Test data for performing verification tasks (Pörn, 1990, p. 24)

| | | | Classical estimates | | |
|---|---|---|---|---|---|
| Component | No. of failures | Operating hours | 5% | MLE | 95% |
| *Other plant* | | | | | |
| 1 | 2 | 17 600 | $2.0 \times 10^{-5}$ | $1.3 \times 10^{-4}$ | $2.6 \times 10^{-4}$ |
| 2 | 1 | 17 600 | – | $5.7 \times 10^{-5}$ | $1.7 \times 10^{-4}$ |
| 3 | 3 | 10 700 | $7.7 \times 10^{-5}$ | $2.8 \times 10^{-4}$ | $6.0 \times 10^{-4}$ |
| 4 | 1 | 10 700 | – | $9.4 \times 10^{-5}$ | $2.8 \times 10^{-4}$ |
| 5 | 0 | 29 500 | – | – | $1.0 \times 10^{-4}$ |
| 6 | 0 | 29 500 | – | – | $1.0 \times 10^{-4}$ |
| 7 | 4 | 15 000 | $1.2 \times 10^{-4}$ | $2.7 \times 10^{-4}$ | $5.2 \times 10^{-4}$ |
| 8 | 1 | 15 000 | $2.4 \times 10^{-5}$ | $1.3 \times 10^{-4}$ | $3.1 \times 10^{-4}$ |
| 9 | 1 | 22 000 | – | $4.6 \times 10^{-5}$ | $1.4 \times 10^{-4}$ |
| 10 | 1 | 22 000 | – | $4.6 \times 10^{-4}$ | $1.4 \times 10^{-4}$ |
| 11 | 0 | 4600 | – | – | $6.5 \times 10^{-4}$ |
| 12 | 0 | 4600 | – | – | $6.5 \times 10^{-4}$ |
| 13 | 1 | 5600 | $2 \times 10^{-5}$ | $1.8 \times 10^{-4}$ | $5.4 \times 10^{-4}$ |
| 14 | 0 | 5600 | – | – | $5.4 \times 10^{-4}$ |
| *Target plant* | | | | | |
| 15 | 3 | 5000 | $1.6 \times 10^{-4}$ | $6.0 \times 10^{-4}$ | $1.3 \times 10^{-3}$ |

**Table 2.3** Pörn's results for data in Table 2.2; $x_{15} = 3, T_{15} = 5000$

| Quantile | $\lambda_{15}$(failures/h) |
|---|---|
| 5% | $8.0 \times 10^{-5}$ |
| 50% | $2.3 \times 10^{-4}$ |
| 95% | $6.3 \times 10^{-4}$ |
| Mean | $2.8 \times 10^{-4}$ |

**Table 2.4** Results for MAPLE and UNICORN, to be compared with Table 2.3

| Quantile | $\lambda_{15}$ | $\lambda_{15}$ | $\lambda_{15}$ | $\lambda_{15}$ |
|---|---|---|---|---|
| | (MAPLE $0.01 < \alpha < 5$, $0.01 < \beta < 50\,000$, $8^3$ gridpoints) | (MAPLE $0.01 < \alpha < 5$, $0.01 < \beta < 50\,000$, $4^3$ gridpoints) | (UNICORN $0.01 < \alpha < 5$, $718 < \beta < 123500$, 2300 samples) | (UNICORN $0.01 < \alpha < 5$, $718 < \beta < 123\,500, 134$ samples) |
| 5% | $8.0 \times 10^{-5}$ | $8.0 \times 10^{-5}$ | $6.1 \times 10^{-5}$ | $7.0 \times 10^{-5}$ |
| 50% | $2.3 \times 10^{-4}$ | $2.1 \times 10^{-4}$ | $1.8 \times 10^{-4}$ | $1.7 \times 10^{-4}$ |
| 95% | $5.7 \times 10^{-4}$ | $4.5 \times 10^{-4}$ | $5.6 \times 10^{-4}$ | $5.9 \times 10^{-4}$ |

**Table 2.5** Classical estimates with 90 per cent confidence bounds and simple Bayes estimates

|  | Maximum likelihood, 90% classical confidence bounds | Simple Bayes, 5%, 50%, 95% quantiles |
|---|---|---|
| Plant 15 alone | $6 \times 10^{-4}[1.6 \times 10^{-4}, 1.3 \times 10^{-3}]$ | $\alpha = 3, \beta = 5000$<br>$1.6 \times 10^{-4}, 5.4 \times 10^{-4}, 1.3 \times 10^{-3}$ |
| All 15 plants pooled | $8.88 \times 10^{-5}[5.6 \times 10^{-5}, 9.8 \times 10^{-5}]$ | $\alpha = 19, \beta = 215\ 000$<br>$5.8 \times 10^{-5}, 8.7 \times 10^{-5}, 1.2 \times 10^{-4}$ |

Task 3 (Section 2.9.5) is to compare results of task 1 with results when ranges of hyperparameters are shifted. The results in Table 2.6 were obtained with MAPLE using the $8 \times 8 \times 8$ grid. The values are considered to represent possible choices according to the Pörn or the simple Bayes heuristic interpretation for hyperparameters.

The median and 95 per cent quantile vary over a factor of 5; the 5 per cent quantile varies over a factor of 3. This is somewhat less variation than that in Table 2.5 between the totally pooled and unpooled data.

Task 4 (Section 2.9.5) is to compare the results of task 1 with the 'uncontaminated model'. The results in Table 2.6 remain virtually unchanged when the contamination parameter $c$ is set equal to zero.

Task 5 (Section 2.9.5) is to compare results of task 1 when dependence is assumed in the hyperprior. As mentioned above, a strong negative correlation might be anticipated between $\nu$ and $\mu'$. The results in Table 2.7 are obtained with UNICORN, using the ranges for $\alpha$ and $\beta$ given in Table 2.4. Results are given for the contaminated and for the uncontaminated model; $\tau(\nu, \mu')$ denotes the rank correlation between $\nu$ and $\mu'$.

**Table 2.6** Values of $\lambda_{15}$ for different truncations of hyperparameters $\alpha$, $\beta$

|  | 5% | 50% | 95% |
|---|---|---|---|
| $\alpha \in [0.01, 5]$<br>$\beta \in [0.01, 50\ 000]$ | $8.0 \times 10^{-5}$ | $2.3 \times 10^{-4}$ | $5.7 \times 10^{-4}$ |
| $\alpha \in [0.01, 5]$<br>$\beta \in [50000, 100\ 000]$ | $5.0 \times 10^{-5}$ | $1.1 \times 10^{-4}$ | $1.9 \times 10^{-4}$ |
| $\alpha \in [0.01, 0.1]$<br>$\beta \in [0.01, 50\ 000]$ | $1.6 \times 10^{-4}$ | $4.9 \times 10^{-4}$ | $9.2 \times 10^{-4}$ |
| $\alpha \in [1, 50]$<br>$\beta \in [0.01, 500\ 000]$ | $7.0 \times 10^{-5}$ | $1.2 \times 10^{-4}$ | $2.1 \times 10^{-4}$ |

**Table 2.7** Effects of dependence in contaminated and uncontaminated models using data from Table 2.4

| | $\lambda_{15}$ contaminated | | | $\lambda_{15}$ uncontaminated | | |
|---|---|---|---|---|---|---|
| | 5% | 50% | 95% | 5% | 50% | 95% |
| $\tau(\nu,\mu')=-1$ | $8.6\times10^{-5}$ | $3.6\times10^{-4}$ | $9.0\times10^{-4}$ | $1.4\times10^{-4}$ | $4.2\times10^{-4}$ | $9.7\times10^{-4}$ |
| $\tau(\nu,\mu')=-0.9$ | $1.1\times10^{-4}$ | $3.1\times10^{-4}$ | $7.0\times10^{-4}$ | $1.1\times10^{-4}$ | $3.4\times10^{-4}$ | $8.0\times10^{-4}$ |

The negative correlation has the effect of widening the 90 per cent confidence bands. The effect is somewhat greater in the uncontaminated model than in the contaminated model.

Task 6 (Section 2.9.5) is to compare results of task 1 when data from different plants are pooled prior to processing. The data in Table 2.2 could be pooled in different ways. First of all, we might consider plant 15 as a population unto itself. In this case the non-informative hyperpriors are used without updating. Second, since plants 1 and 2 have run for the same number of hours, we may guess that these plants are coupled, e. g. by being located at the same site. The same holds for plants 3 and 4, etc. It would be possible to regard these pairs of plants as having drawn the *same* unknown $\lambda$. In that case we should have seven unknown values of $\lambda$, used to update the last failure rate. Third, we might consider the data from plants 1–14 as being generated by the same failure rate, which, however, is different from the failure rate at plant 15. Finally, we might pool all the data and consider all data as generated by one unknown failure rate. This is like the first case, except that the data from all plants have been pooled. The ranges for $\alpha$ and $\beta$ are those from Table 2.4. The results are identical for the contaminated and uncontaminated cases.

Table 2.8 gives some insight into the effect of updating with data from other plants. As the *number* of other plants is reduced from 14 (Table 2.4) to 7 ($\lambda_{2i}=\lambda_{2i-1}$) to 1 ($\lambda_1=\ldots=\lambda_{14}$), the results approach the $\lambda_{15}$ no-update case, even though the total amount of data is the same. The case $\lambda_{15}$ no-update is similar to the simple Bayes case, except that the 90 per cent

**Table 2.8** Effects of different data pooling; hyperpriors from Table 2.4

| | $\lambda_{15}$ | | | |
|---|---|---|---|---|
| | 5% | 50% | 95% | |
| $\lambda_{15}$ no update | $1.6\times10^{-4}$ | $4.9\times10^{-4}$ | $9.1\times10^{-4}$ | prior $(\alpha,\beta,c)$ |
| $\lambda_{2i}=\lambda_{2i-1}$ | $9.0\times10^{-5}$ | $2.5\times10^{-4}$ | $7.1\times10^{-4}$ | |
| $\lambda_1=\lambda_2=\ldots=\lambda_{14}$ | $1.6\times10^{-4}$ | $4.9\times10^{-4}$ | $9.1\times10^{-4}$ | |
| $\lambda_1=\lambda_2=\ldots=\lambda_{15}$ | $1.0\times10^{-4}$ | $1.5\times10^{-4}$ | $2.2\times10^{-4}$ | |
| $\lambda_{15}$ Table 2.5 (unpooled) | $1.6\times10^{-4}$ | $5.4\times10^{-4}$ | $1.4\times10^{-3}$ | |
| $\lambda_{\text{total}}$ Table 2.5 (pooled) | $5.8\times10^{-5}$ | $8.7\times10^{-5}$ | $1.2\times10^{-4}$ | |

confidence bands are a bit wider owing to the hyperprior over $\alpha$ and $\beta$. However, when we go one step further and consider plant 15 as having the same failure rate as plants $1, \ldots, 14$, then the case resembles the $\lambda_{\text{total}}$ of Table 2.5, except, again, for the broadening due to the hyperpriors.

## 2.12   Conclusions

This section gathers some overall conclusions of phase 1 and phase 2 of this research. Most importantly, the T-Book methodology has been validated, from both a mathematical and a computational standpoint. It represents perhaps the most sophisticated implementation of current reliability data base design concepts.

On the other hand, it also inherits shortcomings in the current reliability data base design concepts. In particular, it ignores information from competing failure modes, preventive maintenance, degraded and incipient failure. It also ignores trends, ageing, evidence of maintenance-induced failures and repair times (necessary for availability calculation). Finally, it hard-wires one particular data-pooling scheme.

Qualitative, graphic tools could be used for exploratory data analysis. Such analyses could precede and direct the application of sophisticated mathematical models.

## References

Apostolakis, G. 1982: Data analysis in risk assessment. *Nuclear Engineering and Design* **71**, 375–81.

Apostolakis, G., Kaplan, S., Garrick, B. and Duphily, R. 1980: Data specialization for plant specific risk studies. *Nuclear Engineering and Design* **56**, 321–9.

Box, G.E.P. and Tiao, G.C. 1973: *Bayesian inference in statistical analysis*. New York: Wiley.

Cooke, R.M. 1996: The design of reliability data bases, part I and part II. *Reliability Engineering and System Safety* **51**(2), 137–47, 209–25.

Cooke, R., Bedford, T., Meilijson, I. and Meester, L. 1993: Design of reliability data bases for aerospace applications. Report to the European Space Agency, *Department of Mathematics Report 93-110*, Delft.

Cooke, R., Dorrepaal, J. and Bedford, T. 1995: Review of SKI data processing methodology, *SKI 95:2*. SKI.

Hempel, C. 1968: Maximal specificity and lawlikeness in probabilistic explanation. *Philosophy of Science* **35**, 116–33.

Hora, S.C. and Iman, R.L. 1990: Bayesian modeling of initiating event frequencies at nuclear power plants. *Risk Analysis* **10**(1), 102–9.

IEEE 1984: *IEEE Standard 500-1984*. Piscataway, NJ.

Kaplan, S. 1983: On a 'two-stage' Bayesian procedure for determining failure rates from experimental data. *IEEE Transactions on Power Applications and Systems* **PAS-102**, 195–202.

Kingman, J. 1993: *Poisson processes*. Oxford: Clarendon Press.

Mosleh, A. and Apostolakis, G. 1985: The development of a generic data base for failure rates. *Proceedings of the ANS/ENS International Topical Meeting on*

*Probabilistic Safety Methods and Applications*, San Francisco, CA, 24 February–1 March.

Paulsen, J., Dorrepaal, J., Hokstadt, P. and Cooke, R. 1996: *The design and use of reliability data base with analysis tool, RISØ-R-896(EN)*. Roskilde, Denmark: Risø National Laboratory.

Peterson, A.V. 1977: Expressing the Kaplan-Meier estimator as a function of empirical subsurvival functions. *JASA* **72**(360), 854–8.

Pörn, K. 1990: *On empirical Bayesian inference applied to Poisson probability models*. Linköping Studies in Science and Technology, Dissertation No. 234, Linköping.

Tsiatis, A. 1975: A nonidentifiability aspect of the problem of competing risks. *Proceedings of the National Academy of Sciences USA* **72**(1), 20–2.

## Further reading

American Institute of Chemical Engineers 1989: *Guidelines for process equipment reliability data*. New York: Center for Chemical Process Safety of the American Institute of Chemical Engineers.

Cooke, R.M. 1993: The total time on test statistic and age-dependent censoring. *Stat. and Prob. Let.* **18**(5), 307–12.

Cox, D.R. 1959: The analysis of exponentially distributed life-times with two types of failure. *Journal of the Royal Statistical Society* B **21**, 414–21.

Cox, D.R. and Lewis, P. 1966: *The statistical analysis of series of events*. London: Methuen.

EIREDA European Industry Reliability Data Handbook, 1991: C.E.C.- J.R.C./ISEI 21020 ISPRA (Varese) Italy, EDF–DER/SPT, 93206 Saint Denis (Paris), France.

Green, A.E. and Bourne, A.J. 1972: *Reliability technology*. London: Wiley.

Hora, S.C. and Iman, R.L. 1987: Bayesian analysis of learning in risk analyses. *Technometrics* **29**, 221–8.

IAEA 1988: Component reliability data for use in probabilistic safety assessment, *IAEA TECHDOC 478*. Vienna: IAEA.

Kelly, D. and Seth, S. 1993: *Data review methodology and implementation procedure for probabilistic safety assessments, MTR-93W0000153, HSK-AN-2602*. McLean, Virginia: MITRE Corporation.

Nyman, R. 1995: T-Book seminar 1995-01-27 in Stockholm, Swedish Nuclear Power Inspectorate (SKI). SKI/RA-001/95.

OREDA Committee 1984: *Offshore reliability data*. Norway: Hovik.

Paulsen, J. and Cooke, R. 1994: Concepts for measuring maintenance performance and methods for analyzing competing failure modes. *Proceedings ESREL 94; 9th International Conference on Reliability and Maintainability*, La Baule, France.

Paulsen, J. and Cooke, R. 1994: Analysis of 221 reports using subsurvival functions. *RISOE-I-795(EN)*, Roskilde (to appear in *Reliability Engineering and System Safety*).

Peterson, A.V. 1976: Bounds for a joint distribution function with fixed subdistribution functions: Application to competing risks. *Proceedings of the National Academy of Sciences USA* **73**(1), 11–13.

T-Book Reliability Data of Components in Nordic Nuclear Power Plants, ATV office, Vattenfall AB, S-162 87, Vallingby, Sweden. SKI.

Tomic, B. 1993: Multi-purpose in-plant data system. IAEA, June.

# 3 Bayes' theorem and decision support in 'front line' clinical medicine

F T de Dombal,[1] S E Clamp, M Chan

## 3.1 Introduction

Medical diagnosis and decision making have been around since the time of Hippocrates. Indeed, both are implicit in the Hippocratic oath 'Wheresoever I enter in, there will I do good'. The term 'medix' itself is even older, an Etruscan word implying a village magistrate or decision maker.

However, under the influence of clinicians, especially pathologists from central Europe, until the end of the 19th century clinical medicine and diagnosis was largely qualitative – an art dominated by description and determinism. It was only in the 1950s, partly due to the advent at that time of the electronic digital computer, that the possibilities of probabilist diagnosis began to be explored, the pioneers in this field being Robert S. Ledley and Lee B. Lusted (Ledley and Lusted, 1959). In this new concept of probabilist medicine – possibly aided by computer – it seemed only natural to apply Bayes' theorem; and it is true to say that in the intervening years the vast majority of computer-aided diagnosis and decision support experiments in clinical medicine have been conducted using a Bayesian model for computer prediction.

Despite some successes, however, this concept has yet to be applied in a widespread manner in any area of clinical medicine. In this chapter, therefore, we shall explore the case study of Bayesian analysis in clinical medicine, looking first at the definitions both of decision making in front line clinical medicine and indeed of Bayes' theorem (as applied in these experiments). From a description of one practical exercise in some detail, we shall consider both the benefits and the limitations of Bayesian analysis in clinical medicine. Finally, and perhaps most important of all, we shall consider two further questions: (1) is Bayesian analysis acceptable to those who actually practice clinical medicine, and (2) is there any role for

[1] Deceased.

Bayesian analysis in clinical medicine in terms of decision support in particular?

## 3.2   What do we mean by diagnosis and decision making in 'front line' clinical medicine?

Part of the problem in the introduction of Bayesian analysis into 'front line' clinical medicine has been a lack of understanding of what actually happens in this particular field. This, as Heathfield and Wyatt (1993) have pointed out, has led to a mismatch between user requirements and what has actually been provided. In this section, therefore, we begin by defining some clinical terms related to the application area under study.

'Front line' clinical medicine refers to that part of clinical medicine in which the healthcare provider (usually the doctor but sometimes a nurse or other healthcare worker) interacts directly with the patient. This application thus covers areas such as surgery, paediatrics or obstetrics, but does not refer to areas of clinical medicine such as pathology or pharmacology where the doctor/patient interaction is less direct.

Next, and crucially, it is important to separate out two concepts – first the 'art' of diagnosis, and second the diagnostic and decision making process. This distinction is crucial, not merely because of its clinical importance, but because it has given rise to enormous misconceptions as to what Bayesian analysis and computers can and cannot do. For example, during the 1950s and 60s there was much talk of 'computers doing diagnosis' and although this was subsequently toned down into 'computer-aided diagnosis' in the 1970s, this still failed to take into account that what mattered to doctor and patient is not diagnosis *per se* (in the sense of tying a label on a patient) but the decision making process as a whole (arriving at an optimal line of management in order to solve the patient's problem).

Perhaps the most helpful way to illustrate this crucial difference is set out in Fig. 3.1, which outlines a consensus view of the medical decision making process as it emerged in the 1970s – a view which is still substantially accepted today. In order to arrive at an appropriate decision, the doctor first elicits information from the patient by interview, by physical examination and nowadays often by special tests or investigations. Next, the doctor analyses this information – is it true?; what is the patient's main problem?, what is the most likely cause?; what are other possible causes to be excluded?; and so on. Only then does the doctor make a decision about treatment. As illustrated in Fig. 3.1, even then the decision which the doctor makes may not be a therapeutic one; it may simply be to seek more information, so that the process becomes iterative.

There are many who would argue that even Fig. 3.1 is a highly simplistic description of what is in reality an extremely complex process. For example, the question 'are the data true?' covers a range of problems – from the reliability of observations such as physical clinical findings to the veracity of the patient. Nevertheless, Fig. 3.1 illustrates that talk of 'Bayesian diagnosis'

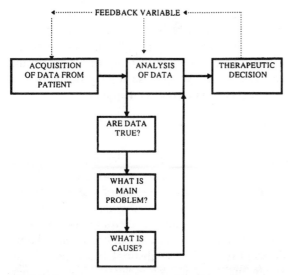

**Fig. 3.1** Simple consensus concept of the diagnostic process

in clinical medicine is wildly simplistic, and if Bayes is to have any application in clinical medicine at all then we must consider how it is to be used, and where it is to fit into the scheme laid out in Fig. 3.1. Before we can do either of these things, however, we need to consider what we mean by Bayes' theorem as it is applied in clinical medicine.

## 3.3 What is Bayes' theorem?

The reader may be astonished at this juncture to find such a question posed, but the question 'what do we mean by Bayes' theorem?' is perhaps not as facetious as it seems – for two reasons, one pedantic, the other intensely practical. Pedantically, one of the authors (FTdeD) has long argued that Bayes' theorem was not written by the Rev. Thomas Bayes at all, but by Richard Price. Far more importantly, use of Bayes' theorem in this application area has involved a number of modifications of that originally proposed – so much so that it is arguable that 'Bayes' theorem' as applied to clinical decision making departs far from the original concept of what Bayes (or Price!) actually intended. Some examples follow.

### 3.3.1 Use of prior probability

Prior probability is an inherent part of Bayes' theorem as it is often stated. Argument rages as to whether Bayes did or did not recognize the need for prior probability; yet many applications in clinical medicine have completely failed to take prior probability into account. Rather, a type of 'implicit prior probability' has been used – in the sense that Bayes' theorem has been applied to specific areas of clinical medicine and has been used to select between a relatively small number of common conditions. For example, in acute abdominal pain (with which we shall deal later), most experiments have dealt with selection between eight or nine common diagnoses. This has implied an inherent prior probability selection (in that rare diseases have not been considered), yet precise prior probabilities, as envisaged by some exponents of Bayes, have not been applied.

### 3.3.2    Treatment of zeros

The next problem which occurs in the use of the Bayesian theory in clinical medicine concerns the treatment of situations in which conditional probability is zero. Clearly, in a multiplicative version of Bayes' theorem, precise application of this criterion would imply that whenever a symptom (out of the relevant cluster of symptoms) is absent, then the disease cannot be present. However, any practitioner of clinical medicine knows immediately that such a situation is totally unrealistic. Clearly zero cannot be inserted into Bayes' formula, and there is even today no clear agreement as to what value should be substituted.

### 3.3.3    Dependence between conditional probabilities

This problem, more than any other, has bedevilled Bayesian analysis in clinical medicine. Unfortunately, it remains a feature of clinical medical data that the symptoms and signs on which a probabilist diagnosis is based are themselves 'dependent' – that is to say where one sign is present, the likelihood of several others is either increased or decreased.

To make matters worse, the dependence is by no means consistent or even reciprocal. For example, a patient who complains of nausea may or may not complain also of vomiting. However, a patient who complains of having vomited has usually (but not always) experienced nausea at that time. These two symptoms are collectively linked to a loss of patient appetite, once again in a way that is by no means straightforward.

There have been many and various attempts to get round this problem. The obvious solution is to compartmentalize the symptoms – i.e. to regard nausea without vomiting, vomiting without nausea, nausea plus vomiting and absence of both symptoms as a four-way choice – but for the reasons outlined above this becomes extremely complex.

Moreover, despite such complicated variants of Bayes' theorem (some of which have required mainframe computers and a substantial amount of time to compute a calculation), it remains to be shown that any of the more complicated variants produce a higher degree of accuracy with a greater degree of consistency than 'simpleton's Bayes'.

For these reasons, most exponents of Bayes' theorem in clinical medicine have ignored prior probability (except in its crudest sense), have substituted a small figure (possibly 0.1) for zero in the calculation, and have ignored dependence of symptoms and signs. This 'simpleton's Bayes' has some merit in that it is relatively easy to calculate and it is this version which has been largely used in clinical medicine to date.

## 3.4 A practical example of Bayesian analysis

Next, it is appropriate to describe how Bayes' theorem has been used in real life in the assistance of diagnosis and decision making in clinical medicine. The clinical area chosen is perhaps the best worked out of all clinical areas, mainly acute abdominal pain on which studies have been carried out for at least 30 years. For clarity, we shall describe in this section the modus operandi for the use of Bayes' analysis in what have become the classical experiments – principally because this was the simplest usage applied – and will describe in further sections the modifications which have been made to this classical usage of Bayes' theorem over the years since the late 1960s.

What follows, therefore, corresponds roughly to the modus operandi described in detail by de Dombal (1979) with only minor adaptations. We will imagine the arrival of the patient suffering from acute abdominal pain in the emergency room of a large general hospital. Under these circumstances the following various steps take place.

*Step 1*  The patient arrives in hospital complaining of abdominal pain of less than 1 week's duration and is admitted to the surgical wards. The patient is seen, interviewed and examined by a number of individuals, including the surgical resident. The surgical resident is then asked to record, either directly or by giving the information to a member of the computing team, the data elicited from the patient (Fig. 3.2). The categories of information that he/she may choose are pre-specified, and a sample of these categories is shown in Table 3.1. This ends for the time being the doctor's participation in the system.

*Step 2*  The data are then entered into the computer using a pre-specified code (Table 3.1). The computer systems used have been described in detail elsewhere (de Dombal, 1984), as have methods of data entry.

*Step 3*  The new case data are first checked for accuracy. The computer then compares details of the new case with details of several thousand similar cases of acute abdominal pain (Table 3.2), calculates the probability of various diagnoses and makes a diagnostic prediction. This calculation derives from the work of Bayes. The resultant data are then displayed to the doctor. Fig. 3.3 shows an outline diagram of this computer analysis.

*Step 4*  The information displayed to the doctor is shown in Fig. 3.4. The symptoms and signs obtained on the history and physical examination are listed. This is a useful check on the correctness of the data entered. Next the

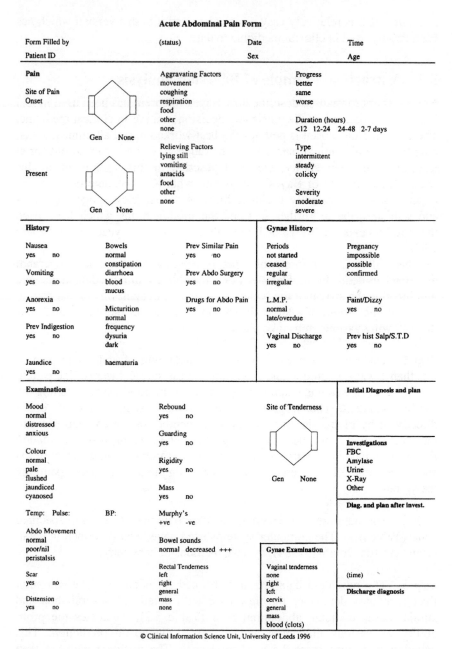

**Acute Abdominal Pain Form**

| Form Filled by | (status) | Date | Time |
| Patient ID | | Sex | Age |

**Pain**

Site of Pain
Onset

Gen    None

Present

Gen    None

Aggravating Factors
movement
coughing
respiration
food
other
none

Relieving Factors
lying still
vomiting
antacids
food
other
none

Progress
better
same
worse

Duration (hours)
<12  12-24  24-48  2-7 days

Type
intermittent
steady
colicky

Severity
moderate
severe

**History**

| Nausea | Bowels | Prev Similar Pain |
| yes    no | normal | yes    no |
| | constipation | |
| Vomiting | diarrhoea | Prev Abdo Surgery |
| yes    no | blood | yes    no |
| | mucus | |
| Anorexia | Micturition | Drugs for Abdo Pain |
| yes    no | normal | yes    no |
| Prev Indigestion | frequency | |
| yes    no | dysuria | |
| | dark | |
| Jaundice | haematuria | |
| yes    no | | |

**Gynae History**

| Periods | Pregnancy |
| not started | impossible |
| ceased | possible |
| regular | confirmed |
| irregular | |
| L.M.P. | Faint/Dizzy |
| normal | yes    no |
| late/overdue | |
| Vaginal Discharge | Prev hist Salp/S.T.D |
| yes    no | yes    no |

**Examination**

| Mood | Rebound | Site of Tenderness |
| normal | yes    no | |
| distressed | | |
| anxious | Guarding | |
| | yes    no | |
| Colour | Rigidity | |
| normal | yes    no | |
| pale | | Gen    None |
| flushed | Mass | |
| jaundiced | yes    no | |
| cyanosed | | |
| Temp:  Pulse:    BP: | Murphy's | |
| | +ve    -ve | |
| Abdo Movement | | |
| normal | Bowel sounds | |
| poor/nil | normal  decreased  +++ | **Gynae Examination** |
| peristalsis | | |
| | Rectal Tenderness | Vaginal tenderness |
| Scar | left | none |
| yes    no | right | right |
| | general | left |
| Distension | mass | cervix |
| yes    no | none | general |
| | | mass |
| | | blood (clots) |

**Initial Diagnosis and plan**

**Investigations**
FBC
Amylase
Urine
X-Ray
Other

**Diag. and plan after Invest.**

(time)

**Discharge diagnosis**

**Fig. 3.2** Data collection form for patients with acute abdominal pain using predefined categories of information

computer makes its diagnostic prediction, given the information already displayed. In computer diagnosis, the system would scan every possible cause of acute abdominal pain and select from its files the most likely cause. This, for reasons already specified, is not possible. The computer therefore

**Table 3.1**  Sample of categories (pre-specified) doctor can choose to describe symptoms of abdominal pain to computer[a]

| Patient characteristics | Category | Patient characteristics | Category |
|---|---|---|---|
| *Sex* | | *Site of pain at present* | |
| Male | 01 | Right upper quadrant | 24 |
| Female | 02 | Left upper quadrant | 25 |
| | | Right lower quadrant | 26 |
| *Age* | | Left lower quadrant | 27 |
| 0–9 | 03 | Upper half | 28 |
| 10–19 | 04 | Lower half | 29 |
| 20–29 | 05 | Right half | 30 |
| 30–39 | 06 | Left half | 31 |
| 40–49 | 07 | Central | 32 |
| 50–59 | 08 | General | 33 |
| 60–69 | 09 | Right loin | 34 |
| 70+ | 10 | Left loin | 35 |
| | | No pain | 36 |
| *Site of pain at onset* | | | |
| Right upper quadrant | 11 | *Aggravating factors* | |
| Left upper quadrant | 12 | Movement | 37 |
| Right lower quadrant | 13 | Coughing | 38 |
| Left lower quadrant | 14 | Respiration | 39 |
| Upper half | 15 | Food | 40 |
| Lower half | 16 | Other | 41 |
| Right half | 17 | Nil | 42 |
| Left half | 18 | | |
| Central | 19 | *Relieving factors* | |
| General | 20 | Lying still | 43 |
| Right loin | 21 | Vomiting | 44 |
| Left loin | 22 | Antacids | 45 |
| No pain | 23 | Food | 46 |
| | | Other | 47 |
| | | Nil | 48 |

[a] Code figures are those used for entering data into the computer.

undertakes only the third diagnostic function in Fig. 3.1 (discriminating among the more common causes for the patient's problem), and the print-out implies that if the patient has one of the more common causes of acute abdominal pain, the most likely cause is as specified.

*Step 5*  This completes the information given to the doctor. The doctor then takes this print-out and uses it in a similar way to a report from any other department (bacteriology, cardiology, biochemistry) to confirm or refute his/her own diagnostic impression and to help decide on a suitable course of action for the individual patient.

**Table 3.2** Example of data base of clinical information about patients with acute abdominal pain entered into computer system prior to commencing computer-aided diagnostic studies

| | Diagnostics[a] | | | | | | |
|---|---|---|---|---|---|---|---|
| Patient characteristics | Appendicitis (100) | Diverticular disease (100) | Perforated peptic ulcer (100) | Non-specific abdominal pain (100) | Cholecystitis (100) | Small bowel obstruction (50) | Pancreatitis (50) |
| *Sex* | | | | | | | |
| Male | 51 | 38 | 79 | 33 | 27 | 54 | 48 |
| Female | 49 | 62 | 21 | 67 | 73 | 46 | 52 |
| *Age* | | | | | | | |
| 0–9 | 19 | 0 | 0 | 20 | 0 | 2 | 0 |
| 10–19 | 31 | 0 | 2 | 34 | 1 | 4 | 2 |
| 20–29 | 22 | 1 | 10 | 16 | 8 | 6 | 4 |
| 30–39 | 10 | 3 | 27 | 8 | 19 | 14 | 16 |
| 40–49 | 6 | 12 | 30 | 8 | 31 | 22 | 28 |
| 50–59 | 5 | 26 | 17 | 7 | 22 | 24 | 22 |
| 60–69 | 4 | 32 | 9 | 4 | 12 | 18 | 20 |
| 70+ | 3 | 26 | 5 | 3 | 7 | 12 | 8 |

*Site of pain at onset*

| | | | | | | |
|---|---|---|---|---|---|---|
| Right upper quadrant | 1 | 8 | 3 | 45 | 2 | 6 |
| Left upper quadrant | 0 | 5 | 1 | 1 | 2 | 6 |
| Right lower quadrant | 23 | 1 | 28 | 3 | 2 | 0 |
| Left lower quadrant | 1 | 0 | 4 | 0 | 2 | 0 |
| Upper half | 4 | 46 | 9 | 27 | 18 | 62 |
| Lower half | 11 | 1 | 10 | 1 | 14 | 0 |
| Right half | 9 | 8 | 5 | 13 | 2 | 2 |
| Left half | 1 | 2 | 3 | 0 | 2 | 2 |
| Central | 44 | 8 | 29 | 6 | 44 | 12 |
| General | 6 | 20 | 8 | 3 | 12 | 10 |
| Nil | 1 | 1 | 1 | 1 | 1 | 1 |

[a] In brackets: number of patients for each diagnostic group.

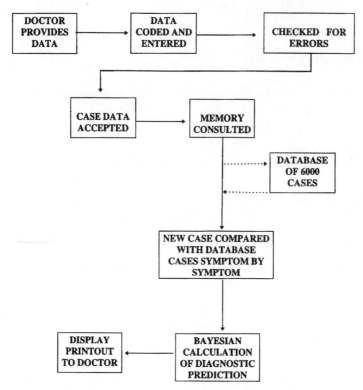

**Fig. 3.3** Outline diagram showing the computer analysis of each new patient's data

## 3.5   Benefits and limitations of Bayesian analysis

### 3.5.1   Benefits of Bayesian analysis

Another problem which has bedevilled the evaluation of computers in clinical medicine has been the failure to define *ab initio* the parameters by which performance is to be assessed.

In acute abdominal pain, the following have been the main parameters put forward as markers of effectiveness of clinical and/or computer performance:

- initial diagnostic accuracy (that of the doctor when the patient first comes to hospital);

- post-investigation accuracy (the accuracy of diagnosis after consultation and investigation – i.e. the point at which the critical clinical decision is made);

- perforation rate (i.e. the proportion of patients with appendicitis whose appendix perforates before the patient is taken to theatre and the appendix is removed);

- negative appendicectomy rates (i.e. the proportion of appendices removed which are normal, indicating the operation was unnecessary);

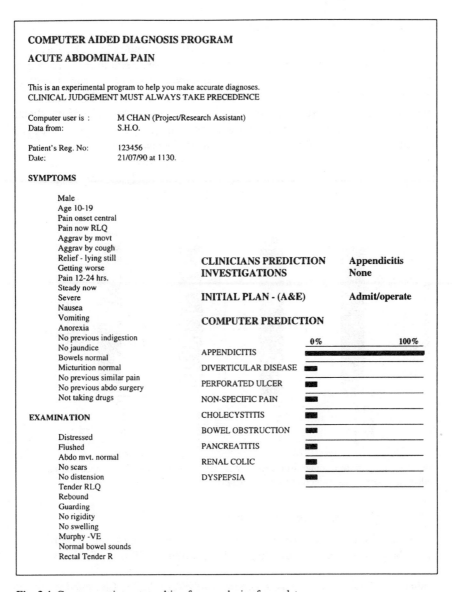

**COMPUTER AIDED DIAGNOSIS PROGRAM**

**ACUTE ABDOMINAL PAIN**

This is an experimental program to help you make accurate diagnoses.
CLINICAL JUDGEMENT MUST ALWAYS TAKE PRECEDENCE

| | |
|---|---|
| Computer user is : | M CHAN (Project/Research Assistant) |
| Data from: | S.H.O. |

| | |
|---|---|
| Patient's Reg. No: | 123456 |
| Date: | 21/07/90 at 1130. |

**SYMPTOMS**

Male
Age 10-19
Pain onset central
Pain now RLQ
Aggrav by movt
Aggrav by cough
Relief - lying still
Getting worse
Pain 12-24 hrs.
Steady now
Severe
Nausea
Vomiting
Anorexia
No previous indigestion
No jaundice
Bowels normal
Micturition normal
No previous similar pain
No previous abdo surgery
Not taking drugs

**EXAMINATION**

Distressed
Flushed
Abdo mvt. normal
No scars
No distension
Tender RLQ
Rebound
Guarding
No rigidity
No swelling
Murphy -VE
Normal bowel sounds
Rectal Tender R

**CLINICIANS PREDICTION**   **Appendicitis**
**INVESTIGATIONS**   **None**

**INITIAL PLAN - (A&E)**   **Admit/operate**

**COMPUTER PREDICTION**

0%                                    100%

APPENDICITIS
DIVERTICULAR DISEASE
PERFORATED ULCER
NON-SPECIFIC PAIN
CHOLECYSTITIS
BOWEL OBSTRUCTION
PANCREATITIS
RENAL COLIC
DYSPEPSIA

**Fig. 3.4** Computer print-out resulting from analysis of case data

- resource utilization (i.e. number of investigations used, number of bed-nights occupied by patients and so on).

Accepting these parameters as markers of performance, there are two aspects relevant to our present discussion. The first is the effect (real or potential) of Bayesian analysis *per se*, and the second is the effect on the performance of clinicians when provided with a computer-aided Bayesian analysis.

### 3.5.1.1   *Bayesian analysis per se*

Fig. 3.5 demonstrates the current position which is quite clear cut. Bayesian analysis of relevant data (in the scheme set out under Section 3.4) provides a more accurate diagnostic forecast than the inexperienced doctor in 'baseline' studies (i.e. totally unaided) – by around 15–20 per cent. However, when the same doctors are provided with structured data collection sheets (necessary for a Bayesian analysis), the impressive difference narrows considerably. Thus, whilst in isolated cases Bayesian analysis has outperformed experienced clinicians, this has not always been so; and it is probably reasonable to conclude that Bayesian analysis *per se* provides much the same accuracy as a reasonably experienced clinician.

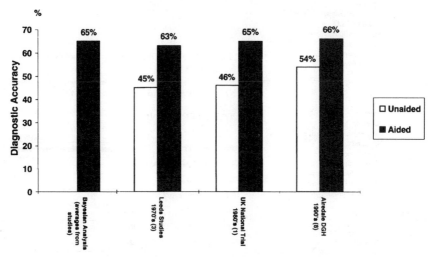

**Fig. 3.5** Comparison of diagnostic accuracy of Bayesian analysis with unaided (baseline) and aided (structured data collection form) clinicians in three major studies in the UK

### 3.5.1.2   *Effects on clinical decision making*

Here, the data are more extensive and are more encouraging. The findings in a large-scale European study (de Dombal, 1993b) (Table 3.3) are

**Table 3.3**   Performance levels in centres involved in the EC Concerted Action in Objective Medical Decision Making. Data involves 14 963 cases

|  | Unaided | Aided (Low–high) | Mean (14 963 cases) |
|---|---|---|---|
| Initial diagnostic accuracy | 48% | (56.7–80.9%) | 65.9% |
| Post-investigation accuracy | 66% | (69.8–96.1%) | 82.6% |
| Perforated appendicitis rate | 27% | (0.0–35.3%) | 13.6% |
| Negative appendicectomy rate | 30% | (1.5–34.2%) | 15.6% |

similar to those observed in a number of other studies and maintained over periods of up to 15–20 years in some hospitals (McAdam *et al.*, 1990). The table indicates an improvement (during time periods when doctors are provided with a Bayesian analysis) of around 20 per cent in diagnostic accuracy and roughly a halving of both perforation and negative appendicectomy rates (with concomitant resource savings which are not detailed in the table).

In summary, therefore, Bayesian analysis *per se* produces similar accuracy of diagnostic performance to that of a senior clinician. Furthermore, when junior doctors are provided with Bayesian analysis, their performance rises to match that of their senior colleagues.

### 3.5.2   Limitations of Bayesian analysis

All of this sounds most encouraging, but there are some severe limitations which have so far prevented more widespread use of Bayesian analysis in this clinical field. The main two are as follows:

- *Restriction of disease categories.* In the example set out in Section 5.4, consideration is restricted to less than 10 disease categories. Certainly these categories account for around 90 per cent of all cases seen; but it is nevertheless greatly irritating to the practising clinician when a favoured diagnosis (which turns out to be correct) is not even considered by a Bayesian analysis because the disease category is not on the list. Thus Bayesian analysis is poor at identifying significant low probability events and identifying rare but important diseases. Attempts have been made to remedy this, but these have been largely unsuccessful due to the difficulty which Bayesian analysis has when faced with conditional probabilities based on small numbers.

- *Sensitivity to varying conditions.* Much more than the clinician, Bayesian analysis is sensitive to varying conditions – such as the clinician's acumen in eliciting signs and/or symptoms, or local variations in an overall pattern. As a result of this, Bayesian predictions tend to be overconfident in clinical medicine, leading to a perfectly reasonable charge of poor calibration. Again, this is unsettling to the practising clinician, who finds it difficult to equate Bayesian confidence (99.99 per cent appendicitis) with the knowledge that the Bayesian system is only 70 per cent accurate!

These two features (along with more mundane considerations such as the length of time taken to use Bayesian systems in an emergency situation) have resulted in the present situation, namely that the 'take-up' of systems demonstrably 20 per cent more accurate than junior clinicians has been patchy to say the least. This must lead to two further questions, which we shall consider in subsequent sections: namely is Bayesian analysis acceptable at all (and if not, why not), and if acceptable, what is the role of Bayesian theory in clinical decision support?

## 3.6   Is Bayesian theory acceptable in medicine?

Put crudely, we may sum up the preceding set of data in the last few sections in two succinct phrases:

• applied Bayesian theory improves inexperienced doctors by 20 per cent;

• doctors will not use it.

This is neither logical nor helpful, and it behoves proponents of Bayes' theorem to enquire why this should be so. Many reasons have been advanced. They range from the reasonable view that doctors do not understand mathematics nor trust mathematics (in which case, how is it that doctors place great store on haemoglobin values?), to the illogical view that human life is too important to be left with a computer (and if one really believes that view, the next time one's airline turns on a final approach and one still cannot see the ground, it would be a good time to get out and walk) (de Dombal, 1993a).

Nevertheless, this is an important issue, and one of the present authors (SEC) has made a particular study of this aspect of Bayesian analysis in clinical medicine. The results of this survey, which involved over 600 subjects (doctors, medical students and members of the public) and was undertaken to examine attitudes to the use of Computer Aided Decision Support Systems (CADSSs) in clinical medicine (Clamp, 1995), revealed that there were substantial reservations concerning the use of these systems. Nearly half of those surveyed did not think that there would be any benefit to diagnosis or decision making, and even more concerns were expressed regarding the effect of using CADSSs on the clinicians' own skills and their relationships with patients.

The reasons for these attitudes are complex. However, one clear factor to emerge was that when doctors did not understand what the computer was doing in making its prediction, this 'black box' effect was seen as a threat to 'clinical expertise'. As the chapters in this book amply demonstrate, it is not easy to explain Bayesian theory to the uninvolved lay person, and lack of understanding of the mathematical basis in CADSSs specifically, and of the decision making process in general, within the medical profession are major problems which need to be confronted if there is to be an increase in the acceptability of such systems in routine clinical care.

## 3.7   Role for Bayesian analysis in clinical decision making

From all of the above, what conclusions can we draw regarding the use of Bayesian analysis in clinical medicine and, in particular, in relation to clinical diagnosis and decision making? As will be apparent from the foregoing remarks, the position is far from static and indeed far from clear – but certain trends can be discerned:

• There is general agreement that the era of 'determinist' medicine is over, and that diagnosis and medical decision making are based (if not on Bayes) at least on probability rather than certainty.

- Bayesian analysis is not capable of dealing with the whole of medicine. Its use in medical decision making is restricted to analysis of data elicited by the clinician, to single clinical areas and problems, and to selection of alternatives between common diseases.

- It follows from this that talk of 'computers doing diagnosis' is absurd. The only possible role for Bayesian analysis is as an adjunct to the efforts of the clinical doctor along the lines described above.

- Even with these limitations, Bayesian analysis will probably remain only part of computer aided decision support systems (as in the examples given). For example, doctors who reject Bayesian prediction still seem to accept the use of Bayes' theorem to select between broad classes, and then to output to the physician what actually happened to patients who fell into this broad class of computer predictions.

- These considerations have a further important corollary. It is unlikely that Bayesian analysis in the near future will outperform the best clinical doctors, but it is highly probable that the addition of a Bayesian analysis to the overall process will result (in selective areas of medicine) in bringing the most junior and inexperienced doctors up to an agreed standard of performance relatively quickly.

- In summary, it is impossible to defend computer diagnosis based on Bayes' theorem as a 'stand-alone' method (independent of the doctor) of making decisions. It is also impossible to defend (for the reasons given) the use of Bayes' theorem on its own, since doctors do not understand percentages or the basis of the mathematics.

On the other hand, however, the imperatives remain (explosion of knowledge, lack of time to acquire it). For these reasons, where Bayes is used as a first stage in a hybrid, evidence-based model of decision support, on the grounds that in several trials it has demonstrably led to improved performance on the part of the clinician, its use is difficult to oppose.

# References

Adams, I.D., Chan, M., Clifford, P.C. *et al*. 1986: Computer-aided diagnosis of abdominal pain: A multi-centre study. *British Medical Journal* **293**, 800–4.

Clamp, S.E. 1995: The impact on and attitudes of society to computer-aided decision support systems in clinical medicine. *PhD thesis*. University of Leeds.

de Dombal, F.T. 1979: Computers and the surgeon – a matter of decision. In Nyhus, L.M. (ed.), *Surgery annual*. New York: Appleton Century Croft, 33–57.

de Dombal, F.T. 1984: Computer-aided diagnosis of acute abdominal pain. The British experience. *Revue d'Epidemiologie et de Santé Publique* **32**, 50–6.

de Dombal, F.T. 1993a: Let me through – I'm a computer. Inaugural Lecture. *University of Leeds Review* **36**, 107–21.

de Dombal, F.T. 1993b: Objective medical decision making. Acute abdominal pain. In Beneken, J.E.W. and Thevenin, V. (eds), *Advances in biomedical engineering*. Amsterdam: IOS Press, 65–87.

Heathfield, H.A. and Wyatt, J. 1993: Philosophies for the design and development of clinical decision support systems. *Methods of Information in Medicine* **32**, 1–8.

Ledley, R.S. and Lusted, L.B. 1959: Reasoning foundations of medical diagnosis. *Science* **130**, 9–21.

McAdam, W.A.F., Brock, B.M., Armitage, T., Davenport, P. and de Dombal, F.T. 1990: Twelve years experience of computer-aided diagnosis in a District General Hospital. *Annals of the Royal College of Surgery* **72**, 140–6.

# 4 Developing a Bayes linear decision support system for a brewery [1]

## M Farrow, M Goldstein, T Spiropoulos

## 4.1 Background

### 4.1.1 Introduction

The objective of the project described in this chapter was to develop a user-friendly computer-based decision support tool for use by managers at the Romford Brewery in Essex. The system was developed with the needs of the managers who would use it very much in mind. At the time of writing, a working system has been developed and tried out with some managers, but since the brewery itself closed, the system is not yet in everyday use. However, discussions have continued about the development and use of the system elsewhere in the brewing industry, and a similar system is being developed with a company which manufactures lawnmowers.

The computer system uses influence diagrams, presented on the computer screen, in the construction, diagnostic checking and use of representations of the beliefs of industrial managers about quantities of interest in the management of their industry. Although the particular industry involved was a brewery and the project was carried out with the cooperation of brewery staff, the basic system is general and can be used similarly in other industries.

Even in an apparently simple part of the brewing process like the packaging plant where beer is put into kegs, there are many variables to be considered. There are several types of beer and sizes of keg. Demand for the beer varies over time. Targets for production are set but are not always met.

[1] This work was carried out with financial assistance from the Polytechnics and Colleges Funding Council and with the cooperation of the Romford Brewery Company Ltd. The language [B/D] was developed by David Wooff and Michael Goldstein under grants from the Science and Engineering Research Council. A version with manual and tutorial guides is freely available over the Internet.

The nature of the brewing process itself requires that decisions which affect the brewery's ability to meet demand must be made several weeks in advance. At any time there are stocks of various beers in various states, orders from depots, sales by depots to the retail trade, production targets, actual production, etc.

Experienced personnel have a great deal of expertise in terms of the prediction of the values of variables and the effects of decisions. However, to make full use of this expertise, it is necessary to link it to the large quantity of information about the changing state of the brewery system which is now available 'on-line' and to make the expertise available to other staff. Thus we need to construct a computer-based description of the beliefs of these experts.

### 4.1.2   What the system is required to do

The computer system must allow managers, most of whom have limited mathematical and statistical backgrounds, to construct and modify coherent representations of their beliefs and uncertainties about a large number of related quantities. It should have model checking and diagnostic tools to help in this process. It must be able to use the beliefs to produce forecasts of future values. As a decision support tool, the system must be able to show the expected effect on future values of circumstances which might occur or actions the managers might take, that is to conduct a 'what if' analysis. It must be possible to enter data from the brewery and elsewhere as they arrive and to revise predictions accordingly. It should be possible to use the comparison of observed and expected values to warn of the need to modify the belief specification. The system should be usable by those experienced managers who construct and modify the belief representations and also by colleagues who merely consult it.

In the context of, for example, the packaging plant at the brewery, managers might require answers to questions such as the following:

- How much stock will we have at the weekend?

- Will we meet orders?

- How much resources should be put to producing each beer?

- Do we need an extra shift?

- Do we need to start building up stocks for Christmas?

- What is likely to happen if ...?

### 4.1.3   The Bayes-linear approach

Provided we could actually develop and analyse such a specification, a standard Bayesian approach, involving a coherent belief specification in the form of a joint probability distribution over all of the unknowns, would enable us to find the answers to questions of interest involving, for example, revised beliefs given information on the values of some of the unknowns. However, there are difficulties with this approach. Firstly, the specification of a complete joint probability distribution is likely to be beyond what we can reasonably expect from most managers, especially in the limited time they have available. Even if we first identify a conditional independence structure, perhaps using an influence diagram, the conditional distributions which are still required will be extremely difficult to specify realistically. Use can be made, of course, of convenient distributional forms, but these are unlikely to be accurate representations of actual beliefs or, for that matter, to be properly understood by industrial managers. Secondly, the computational problems involved in using such a belief specification, for example in evaluating high-dimensional conditional distributions, are likely to be severe.

The system therefore uses the Bayes-linear approach in which only a limited number of prior expectations, variances and covariances, rather than complete probability distributions, need to be specified. These expectations relate directly to the quantities of interest and their associated uncertainties and are therefore more likely to represent genuinely held beliefs. The values are specified directly. The expert gives subjective expectations of unknown quantities, variances which quantify the uncertainties in the values of the unknowns and covariances which describe the associations in the sense that learning the value of one quantity would cause a revision in the expectation of another.

When the values of some quantities become known, the expectations of the others can be adjusted according to the linear rule which minimizes expected quadratic loss. The calculation of adjusted beliefs is analogous to finding conditional expectations but is computationally undemanding. This does not necessarily mean that we are restricted to working with linear functions of the quantities of interest since we can include in the prior specification any function of any unknown as another unknown, provided we are willing to specify the required mean, variance and covariances.

Consider two vectors of quantities, $\underline{x} = (x_1, \ldots, x_n)'$ and $\underline{y} = (y_1, \ldots, y_n)'$. The prior mean vectors for these are $\underline{\mu}_{0x}$ and $\underline{\mu}_{0y}$. The prior variance matrices for $\underline{x}$ and $\underline{y}$ are $V_{0xx}$ and $V_{0yy}$ such that the $ij$ element of $V_{0xx}$ is the covariance of $x_i$ and $x_j$ (or the variance of $x_i$ if $i = j$) and similarly for $V_{0yy}$. The covariances between $\underline{x}$ and $\underline{y}$ are in the $m \times n$ matrix $V_{0xy}$ the $ij$ element of which is the covariance of $x_i$ and $y_j$. Let $V_{0yx} = V'_{0xy}$.

Suppose now that we observe the value of $\underline{x}$. The *adjusted* expectation of $\underline{y}$, given this information, is then

$$\underline{\mu}_{1y} = \underline{\mu}_{0y} + V_{0yx}V_{0xx}^{-1}(\underline{x} - \underline{\mu}_{0x}) \qquad (4.1)$$

and the adjusted variance of $\underline{y}$ is

$$V_{1yy} = V_{0yy} - V_{0yx}V_{0xx}^{-1}V_{0xy} \qquad (4.2)$$

Equation (4.1) is the linear rule in the elements of $\underline{x}$ which minimizes our prior expectation of $(\mu_{1yi} - y_i)^2$ for $i = 1, \ldots, m$, where $\underline{\mu}_{1y} = (\mu_{1y1}, \ldots, \mu_{1ym})'$, and $V_{1yy}$, as given by eq. (4.2), is our prior expectation of $(\underline{\mu}_{1y} - \underline{y})(\underline{\mu}_{1y} - \underline{y})'$. If $V_{0xx}$ is not of full rank then we may use the Moore–Penrose generalized inverse in the above equations.

Even with only the first two moments required, making a genuine and coherent belief specification for a set of unknowns with complex interrelationships may not be easy, especially for those whose skills lie in other areas. We therefore develop the qualitative aspects of the beliefs of the expert by graphical modelling. Smith (1989) has shown that influence diagrams can be used to represent general relationships such as Bayes-linear structures where the usual conditional independence of probabilistic structures is replaced by weak conditional independence. Thus our influence diagrams differ from the more usual type. Each node represents a vector of quantities. In the brewery, for example, there could be an element of the vector for each of the beers made there. Consider Fig. 4.1. In the usual type

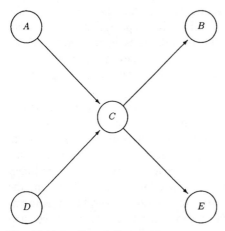

**Fig. 4.1** Bayes-linear influence diagram

of influence diagram, the quantities represented by nodes B and E would be conditionally independent given the value at C. Our interpretation of the diagram is that, in our adjusted beliefs, after linear fitting on the value of the vector at C, the correlations between the quantities at B and those at E are zero. Also, once we know C we would regard A and D as irrelevant to the linear prediction of B and E.

Goldstein (1990) discusses diagrams of this type and points out that the arcs may be labelled with coefficients to denote the strengths of relationships. In our diagrams, we label arcs with coefficients which represent strengths of influence in a way that we shall discuss in Section 4.3.2.

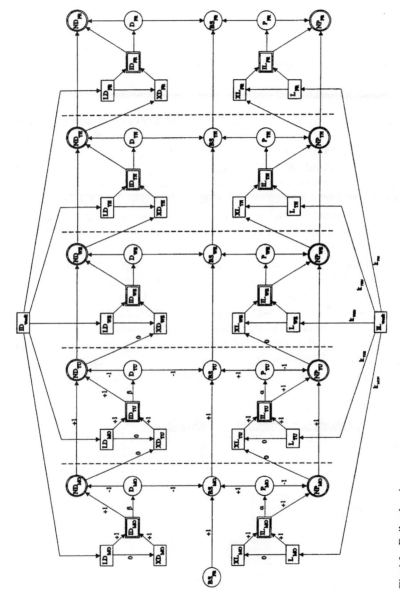

**Fig 4.2** Daily planning

Fig. 4.2 illustrates the complexity which may be involved. This merely shows part of a larger structure representing beliefs about the brewery packaging plant. Nodes representing some quantities relevant to the packaging plant are shown for the five working days of a week. For example, nodes $P_{MO}$, $P_{TU}$, ..., $P_{FR}$ are the productions on Monday, Tuesday, ..., Friday. Node $BS_{MO}$ is the brewery closing stock on Monday and so on.

## 4.2 The computer system 'Foresight'

### 4.2.1 Introduction

The project with the brewery involved the development of a computer system, called 'Foresight', for building the qualitative and quantitative structure of problems in the form of Bayes-linear influence diagrams. Some general ideas about such model building are discussed in Farrow and Spiropoulos (1992). Once the expert's beliefs have been elicited in this way, it becomes possible for managers to use the program to make predictions, to help in decision making or to do 'what if' analysis. As data become available, the beliefs can be updated automatically by the program.

Foresight uses a graphical interface for the development of Bayes-linear influence diagrams. The qualitative structure of the diagram is built up, then quantitative information, such as prior means and variances, can be added. Fig. 4.3 shows a Foresight screen during the building of the qualitative structure shown in Fig. 4.19, which represents part of the structure shown in Fig. 4.2. One of the nodes is being repositioned on the screen.

Foresight has been designed in an object-oriented way and it runs on PCs in the Microsoft Windows environment. Technical details about the implementation are presented in Spiropoulos (1995). The development of

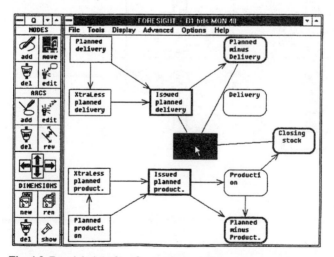

**Fig. 4.3** Foresight interface for moving a node

the qualitative structure of Bayes-linear influence diagrams is presented in the remainder of Section 4.2, and the interface for the specification of quantitative prior information is described in Section 4.4.

We discuss Foresight partly as a program of interest in its own right, but also as an illustration of the potential for general Bayes-linear interfaces for decision support.

### 4.2.2 Node types and operations in Foresight

A Bayes-linear influence diagram (BLID) is a graphical interpretation of a partially specified belief structure using a directed acyclic graph that consists of nodes and arcs. Each random quantity of the belief structure is represented by a node and is assigned a prior mean and a variance. (More generally, a node may represent a vector of quantities, but for simplicity here we discuss scalar quantities. In the current implementation of Foresight, vector quantities are built up by copying the structure in 'layers' using the 'Dimensions' facility: see Section 4.2.4.) In general, the variance assigned to each node can be split into two parts. The first part represents the portion of the variance that can be explained by the uncertainty in its predecessor nodes. The second part represents the residual variation of the node when it has been adjusted by all of its predecessor nodes. This is referred to as the *specific* variance. The notion of specific variance is important to the computer implementation because it is the main criterion used to determine whether a model is internally coherent at the quantitative elicitation stage. Foresight calculates specific variances automatically and it informs the user when negative values are detected. There are different types of nodes represented in Foresight, but for the purposes of this chapter, only the following nodes representing random quantities are considered:

- A *chance* node is assigned a prior mean and a variance and has a positive specific variance, indicating that, if all the predecessors were known, there would still be some uncertainty left in the quantity represented by the chance node. It is represented graphically by a circle or an ellipse. A special type of a chance node is a *common uncertainty node*. Common uncertainty nodes represent non-observable underlying common factors which quantify the uncertainty that is common to two or more quantities. They are assigned variances, but their prior means are always zero. They are used mainly to explain correlations between nodes.

- When a chance node represents a quantity that is under the decision maker's control, it is called a *decision* node. As with chance nodes, decision nodes are assigned a mean and a variance, but they are represented graphically by a rectangle. For example, the quantity 'Monday's planned production' is always under the operation manager's control and can be represented by a decision node. Also, since planned production is not the same every Monday, a variance is assigned to the node as well.

- *Deterministic* nodes represent quantities whose value is a deterministic linear function of the values of their predecessor nodes, so that the specific variance of a deterministic node is always zero. A deterministic node is represented graphically by a double circle or a double ellipse. A deterministic node can also be used to represent a decision that depends deterministically and linearly on its predecessor nodes. In this case, the decision is made according to a chosen linear rule, and the corresponding node is referred to as a *deterministic decision* node. Deterministic decision nodes are represented graphically by a double rectangle, to distinguish them from ordinary decision and deterministic nodes, and their specific variances are zero.

In Foresight, nodes can be created by clicking on the *add* icon of the *nodes* group shown in Fig. 4.4. The user can type the name of the node in the box and select a node type by clicking on the appropriate option displayed at the bottom of the window. The position of a node on the screen can be changed by clicking on the *move* icon and then dragging the node to the new coordinates. The name and type of a node can be changed at a later stage using the *edit* icon. When a node is deleted (*del* icon), all the arcs associated with that node are deleted as well, and the diagram is reordered automatically. Node deletion in Foresight is used to change the qualitative structure of the diagram and is not related to any node removal operations for evaluating influence diagrams.

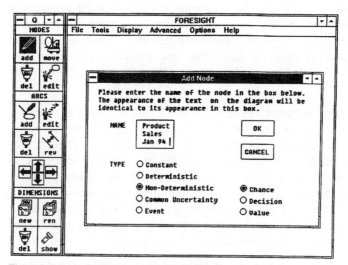

**Fig. 4.4** Foresight interface for adding nodes

### 4.2.3  Foresight interface for arc operations

Arcs entering a node represent dependence without implying causality. Therefore, the absence of arcs between nodes of the diagram implies weak conditional independence, as discussed in Smith (1989). Arcs entering decision nodes are called *informational* and imply time precedence, indicating information available prior to the decision. Arcs have associated with them arc coefficients which are used to calculate the covariance between the nodes. The specification of arc coefficients is discussed in Section 4.4.

Using Foresight, arcs can be created by clicking on the *add* icon of the arcs group (see Fig. 4.5) then on the intended parent node and finally on the

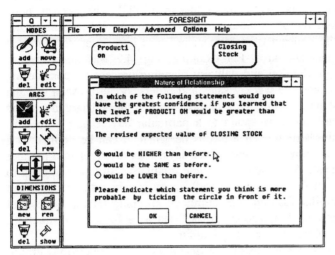

**Fig. 4.5** Foresight interface for arc creation

intended child node. Before the creation of an arc, Foresight checks whether the intended arc is valid by examining the type of the associated nodes and ensuring that directed cycles are not created. There are also various rules forbidding arcs between nodes of certain special types. Such rules are enforced automatically, and when the user intentionally or unintentionally changes the type of a node to one that makes an arc invalid then, before the change is carried out, Foresight informs the user and gives the choice of aborting the operation or continuing and deleting the resulting invalid arc automatically. If the criteria are satisfied, then Foresight presents the user with a window such as that shown in Fig. 4.5, which contains a question designed to elicit the sign of the covariance between the two nodes. For example, considering the nodes 'production' and 'closing stock', the user might click on the 'would be HIGHER than before' statement and thus indicate a positive covariance. The sign of the covariance can be changed later intentionally using the *edit* icon or when the value of the arc coefficient is specified, or unintentionally when Foresight calculates the covariance between the nodes. After the arc is created, the diagram is reordered at run time. Arcs can be deleted or reversed using the *del* and *rev* icons respectively, and such arc operations represent intentional changes to the underlying covariance structure of the diagram. Before completing the arc reversal, Foresight checks whether the operation will create a directed cycle or whether the new arc will be invalid. As in the case of node deletion, the arc reversal operation is not part of a mechanism for evaluating influence diagrams.

### 4.2.4   Foresight dimensions

A detailed account of using influence diagram nodes to represent collections of random quantities is given in Goldstein (1990) where each node represents a random vector. Although in Foresight the vector notation is not used explicitly, the concept of a 'dimension' was introduced to represent collections of nodes and arcs. For example, Fig. 4.2 was developed to support the daily production and delivery planning process by considering five working days, and the weekly sales and orders. It can be seen that the subdiagrams contained within two dashed vertical lines share the same qualitative structure and correspond to a particular day. Using Foresight, when the qualitative structure of a day is designed, it can be defined as a dimension and then copied to create other days, thus resulting in a number of 'time dimensions'. Also, since the same qualitative structure applies to the other products of the company as well, many 'product dimensions' can be created in a similar fashion. Therefore, by grouping related nodes and arcs into subdiagrams, dimensions can be used as a form of abstraction, making it easier to represent complex diagrams.

Dimensions can be created using the *new* icon of the dimensions group, and changes to the dimension's name can be done using the *ren* icon. A whole dimension can be deleted using the *del* icon, and Foresight will reorder the diagram automatically. The icon *show* is used on its own to display different dimensions on the screen, and in combination with the arc icons to create and edit cross-dimension arcs.

## 4.3   Belief modelling

### 4.3.1   Approach to belief modelling

To make use of Foresight in the brewing industry, managers will often wish to construct a representation of beliefs about a large number of quantities where the relationships between these quantities lead to a complex covariance structure. In this section we discuss how to do this. The quantities may include values taken by a variable at a series of time points, such as the amount of a beer in stock at the end of each week, or values of a variable at the same time but for different 'individuals', such as the sales of different beers or the sales of a beer at different depots. Foresight is designed to simplify the construction of such belief representations. However for some complex features, such as the time series of sales of several beers, the help of a specialist analyst is still likely to be required by many managers.

Constructing a coherent covariance structure to represent our actual beliefs would be difficult in the absence of a structured approach. However, we have adopted a general approach which will often make elicitation and belief modelling relatively straightforward.

To illustrate the approach we consider first the simple case where we have a group of exchangeable quantities. These could be the production of a particular beer on each of several days (although, as we shall see below, in practice we might well not regard these as exchangeable). We can represent our beliefs about these in terms of the sum of a mean $M$ and a deviation $\varepsilon$ from that mean. Thus the value of observation $i$ is written as

$$Y_i = M + \varepsilon_i$$

where our expectation of $\varepsilon_i$ is zero and $\varepsilon_i$ is uncorrelated with $M$ and with $\varepsilon_j$ for $i \neq j$. We need to specify an expectation for $M$ and variances for $M$ and $\varepsilon_i$.

Suppose now that we will have observations of two types so that we have two means, $M_1$ and $M_2$. We need two expectations, $E(M_1)$ and $E(M_2)$, two variances and a covariance. Rather than specifying the covariance directly we might regard $M_1 - E(M_1)$ and $M_2 - E(M_2)$ as being made up of linear combinations of (in this case three) zero-mean, mutually uncorrelated quantities which we term 'uncertainty factors'. Thus

$$M_1 - E(M_1) = k_1 E_1 + k_3 U$$
$$M_2 - E(M_2) = k_2 E_2 + k_4 U \tag{4.3}$$

Here $E_1$ and $E_2$ are termed 'specific uncertainty factors' and $U$ is a 'common uncertainty factor' by analogy with factor models. Note that, although $E_1$, $E_2$ and $U$ do not correspond to observables, $k_3 k_4 E(U^2) = E\{[Y_{1i} - E(M_1)][Y_{2i} - E(M_2)]\}$, $k_1^2 E(E_1^2) = E\{[Y_{1i} - E(M_1)][Y_{1j} - E(M_1)]\} - k_3^2 (U^2)$ and $k_2^2 E(E_2^2) = E\{[Y_{2i} - E(M_2)][Y_{2j} - E(M_2)]\} - k_4^2 E(U^2)$, where $Y_{gi}$ is the $i$th observation of type $g$ and $i \neq j$. Thus, for example, by fixing the quantities $k_1, \ldots, k_4$ we can relate the the variances of the uncertainty factors directly to the variances and covariances of observable quantities. The diagram corresponding to the model (eq. 4.3) is shown in Fig. 4.6. The quantities $k_1, \ldots, k_4$ are marked on the respective arcs of the figure and are termed 'arc coefficients'.

In some cases, the two types of observation might refer to the same variable, perhaps measured under different conditions, for example at different depots. In such cases we would usually set $k_1 = k_2 = k_3 = k_4 = 1$, or, for a negative covariance, for example, $k_1 = k_2 = k_3 = 1$ and $k_4 = -1$. Completion of the variance-covariance specification then amounts to assigning variances to $E_1$, $E_2$ and $U$. This may be done by first assigning marginal variances to $M_1$ and $M_2$ then deciding how much is 'shared' between them for the variance of $U$. This seems to be a more appealing approach to the problem for some people than thinking directly in terms of the covariance or correlation. Alternatively, we could ask a question about the reduction in

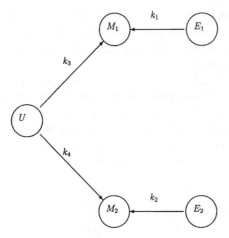

**Fig. 4.6** Two related means

variance of $M_2$ if $M_1$ were to become known since this reduction is $V_{12}^2/V_{11}$ where $V_{11}$ is the variance of $M_1$, and $V_{12}$ is the covariance of $M_1$ and $M_2$.

If $M_1$ and $M_2$ refer, for example, to variables measured on different scales, we might prefer to allow $k_3$ or $k_4$ to take a value other than $\pm 1$ to allow for scale differences. (In fact, we could fix all of the uncertainty factor variances at 1 and just vary the arc coefficients to achieve the desired result.)

When there are more than two quantities involved, a variety of more complex structures are possible. Our approach is first to choose an appropriate qualitative structure, which we can represent by a diagram, then to assign variances to the mutually uncorrelated factors at the 'bottom level' of the diagram. The whole covariance structure is then built up. A first attempt at filling in the variances will not always produce acceptable results and adjustments will have to be made, but this is, in itself, part of the elicitation and modelling process.

Fig. 4.7 shows one possible structure with three means. These could, for example, relate to sales at three depots. Here $U_3$ represents an overall uncertainty in the general level. Closer associations between $M_1$ and $M_2$, and between $M_2$ and $M_3$, are formed by the presence of $U_1$ and $U_2$. Notice in particular that the total number of uncertainty factors is equal to the total number of variances and covariances. We could choose to make the number of uncertainty factors less, and in so doing make a qualitative statement about our beliefs, but we never need more.

Some examples of this sort of modelling in the context of medical experiments are given in Farrow and Goldstein (1992, 1993). The first of these papers also uses influence diagrams.

An example from the brewery project involves the sales of four beers. For the moment we can think of predicting the sales of the beers, perhaps at a number of depots. In fact, as we shall see in Section 4.3.4, the structure described here was used to represent beliefs about a sequence of exchangeable one-step-ahead forecast errors in the beer sales.

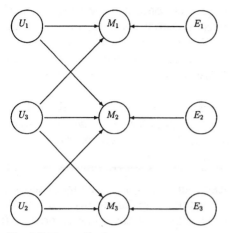

**Fig. 4.7** Three related means

Clearly there are associations in beliefs about the sales of the four beers. The structure shown in Fig. 4.8, with appropriate specific variances and arc coefficients, can represent the 10 observed sample variances and cross covariances from past data remarkably well despite being specified by only seven parameters. The qualitative structure we have used here, by implying some weak conditional independences, has reduced the number of quantitative specifications we need to make. The four nodes $\varepsilon_{1t}, \ldots, \varepsilon_{4t}$ represent the random inputs into the sales of the four beers. Perhaps the most important factor is total beer sales. Therefore node 'Total' represents overall sales. Beer 1 is a bitter while the other three are lagers, so $\varepsilon_{1t}$ is the difference between 'Total' and 'Lager'. Similarly, beers 2 and 3 are thought to compete in the same part of the market for lager while beer 4 has a more distinct clientele, so 'Lager' divides first into $\varepsilon_{4t}$ and 'Q', and then 'Q' is split between $\varepsilon_{2t}$ and $\varepsilon_{3t}$. Each of the nodes $\varepsilon_{1t}, \varepsilon_{2t}, \varepsilon_{4t}$ is given deterministically as the difference between its parents.

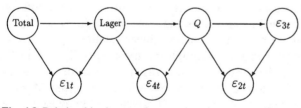

**Fig. 4.8** Relationships between beers

### 4.3.2  Belief modelling for prediction systems

In this project we were particularly concerned with systems where quantities are to be observed at a sequence of time points. In effect, we have a multivariate time series where the number of unknowns at each time point may, in practice, be very large. The project concerned the production of several products in a brewery. In other industrial contexts, there may be hundreds of different product lines and the quantities of each in several locations or stages of production, for example, may be required. Predictions may be required for underlying unobserved quantities as well as for future values of observables.

In such systems there is usually a natural ordering of the quantities. Thus we tend to think of adjusted means and variances for later quantities, given earlier quantities. It is also often the case that the pattern of relationships from one time step (e.g. week) to the next is repeated.

As an example, consider the simple structure shown in Fig. 4.9. Here we have four unknowns and therefore need four means, four variances and six covariances. However we have decided that, given knowledge of the values of $A$ and $D$, we would not use that of $C$ in order to predict $B$ linearly.

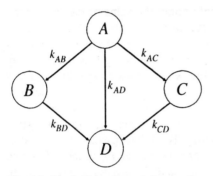

**Fig. 4.9** Simple Bayes-linear influence diagram

Assessment of the means (e.g. $m_A, \ldots, m_D$) should present no particular difficulties. The mean required for a node is the marginal mean. That is the expectation when all other quantities are unknown. This mean is also the Bayes-linear adjusted expectation when all of the direct predecessors are known to take their marginal mean values and none of the direct successors is known. In cases where there are complex linear relationships between means, it might be helpful to work through the mean specification in the direction of the arcs. (In the example, assess $m_A$ first. Then determine the predictions, $m_B$, $m_C$ of $B$ and $C$ given the knowledge only that $A = m_A$. Finally determine the prediction, $m_D$, of $D$ given the knowledge that $A = m_A$, $B = m_B$ and $C = m_C$.)

We model the uncertainty structure, that is the variances and covariances, as follows. The value of a node such as $A$, with no direct predecessors, is given by

$$A = m_A + U_A \tag{4.4}$$

where $U_A$ represents the 'uncertainty' associated with $A$. The values of the other three nodes in the example are as follows:

$$B = m_B + k_{AB}(A - m_A) + U_B$$

$$C = m_C + k_{AC}(A - m_A) + U_C \tag{4.5}$$

$$D = m_D + k_{AD}(A - m_A) + k_{BD}(B - m_B) + k_{CD}(C - m_C) + U_D$$

Each of the four 'uncertainties', $U_A, \ldots, U_D$, has zero mean and is uncorrelated with any other, and we assign to them variances, $v_A, \ldots, v_D$. Notice that, for example, the marginal variance of $B$ is $k_{AB}^2 v_A + v_B$. One way to proceed with the elicitation is to determine the marginal variances first, and then, as the correlations are built up by the assignment of non-zero values to $k$ coefficients, to make any necessary adjustments to the variances of the uncertainties in order to maintain the values of the marginal variances. Alternatively, in some cases, it may be more appropriate to elicit the specific variances (e.g. $v_B$, $v_C$, $v_D$) directly, checking all the time that the implied marginal variances are acceptable. In eliciting any variance it may be found easier to work in terms of standard deviation or, making a suitable judgement as to the shape of an underlying distribution, a probability interval, for example 95 per cent, for the random quantity.

The arc coefficients, i.e. the coefficients $k$ in eqs (4.5), relate the change in expectation for a node to the deviation from prior expectation of a parent node, although, as explained in Section 4.4.4, the relationship is not always straightforward. Two methods of elicitation, for example for $k_{AD}$, can be used to provide a mutual check. Firstly, $k_{AD}$ can be determined by asking for the adjusted expectation of $D$ if it were known that $A = m_A + \delta$, $B = m_B$ and $C = m_C$, where $\delta$ is small enough to keep $A$ within a plausible range but large enough to give a reasonably precise value for $k_{AD}$ from the effect on the expectation for $D$. Secondly, we can ask for the adjusted variance of $D$, given knowledge of the values of $B$ and $C$, and compare this with the specific variance of $D$ to find the value of $k_{AD}$ apart from its sign. The sign is elicited separately (see Section 4.4.4). In both of these elicitation methods, we are assuming that the expert is able to give reponses corresponding to Bayes-linear belief adjustment. That is, we are assuming that $k_{AD}$ is a constant which does not depend on the value of $A$.

Suppose that $A$, $B$ and $C$ in Fig. 4.9 are actually the sales, for some future period, of three brands of beer. If $D$ represented total sales of the three brands, then clearly we would have a deterministic relationship here, with $k_{BD} = k_{AD} = k_{CD} = 1$ and $v_D = 0$. Notice that, even before values for $k_{AB}$ and $k_{AC}$ are determined, we are making a statement about the relationships, in our beliefs, between sales of the three brands in that, if we knew the sales figures for $A$ and $B$ (but not the value of $D$), we would use only that for $A$ to predict the figure for $C$. The model shown in Fig. 4.10 might be more reasonable for this example. Here we introduce a new node $E$ which represents an underlying common uncertainty factor in $A$, $B$ and $C$. This might be regarded as a general beer sales index and could, without loss of generality, be given mean zero and variance 1; then, $A = m_A + U_A + k_{EA}E$

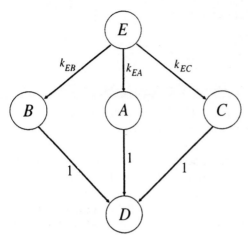

**Fig. 4.10** Sales with a common uncertainty factor

etc. The purpose of *E* is simply to introduce appropriate correlations between *A*, *B* and *C*. A more complex diagram, which was actually used for the four beers, is shown in Fig. 4.8.

In cases where our beliefs about the future behaviour of the system are best represented by a non-stationary model, the means we specify initially for values in the distant future may have little relevance by the time those values occur. However, at the time when we specify our beliefs we can refer to the current values. As time goes on from there, the originally specified means serve as reference values in calculating the continually updated predictions. Alternatively it may be sufficient to specify means for an initial set of nodes and allow other means to be generated by the operation of the arc coefficients. In such non-stationary models, it will often be best to specify specific variances directly rather than marginal variances.

Further discussion of belief modelling for time series is given in Section 4.3.4.

### 4.3.3   Example: Beliefs about packaging

Much of the development of the system was done using the packaging plant as a 'test bed'. Here beer is drawn from vessels in the brewery and packaged into kegs. One of the main concerns in the packaging plant is to meet demand from the distribution depots without holding excessive stock or holding stock for long periods. Since beer making takes several weeks, advance planning is very important.

Fig. 4.2 shows part of a diagram constructed for the packaging plant. The subscripts MO, TU, ..., FR refer to the days of the week. Nodes BS are brewery stock while P stands for production and D for deliveries. Nodes L and LD represent the planned production and deliveries. The remaining nodes refer to the process of adjusting plans on a daily basis, for example to make up shortfalls from previous days' plans. The brewery produced four beers, each packaged into kegs of two sizes. There are thus eight products, so each node can be thought of as an 8-vector. There are various interrelationships between the elements within a vector. In the rest of this chapter, for simplicity, we ignore the two sizes of keg and work in terms of total volumes for the four beers. Thus each node will be a 4-vector. In fact, as explained in Section 4.2.4, this is done in Foresight by creating extra layers of the diagram to represent the 'dimensions' of the vectors.

The nodes with double rings are deterministic in the sense that the specific variance is zero. The rectangular nodes are decision nodes. The arcs into these nodes show what information is used to make the decisions. Some of the arc coefficients are shown in Fig. 4.2. When a linear rule is not given for a decision the coefficients on arcs into the decision node are set at zero.

Filled kegs become part of stock held at the brewery until they are delivered to depots. The depots issue orders in advance to the brewery and these can be used as information in planning the packaging, but for various reasons the deliveries need not exactly match the orders. The depots sell the beer to the retail trade. The concern of the depots is inventory control, so the orders they issue are strongly dependent on sales. Therefore, in the next section, we consider prediction of sales from depots to the retail trade.

### 4.3.4   Stochastic inputs: Time series of sales

There is a strong tradition in time series analysis of using linear sums of past and present observations to forecast future values. There is thus a natural relationship with Bayes-linear methods.

For clarity at this point we discuss belief specification for a univariate series observed at equally spaced time points. The generalization to multivariate series is reasonably straightforward.

Forecasting methods often use the idea that, in some way, the future will resemble the past. In particular, we are often willing to construct a linear predictor, based on observations up to time $t$, to forecast the observations at times $t + 1$, $t + 2$, .... Suppose the predictor for $Y_{t+1}$, the observation at time $t + 1$, is

$$\hat{Y}_t(1) = \alpha + \sum_{i=0}^{\infty} \beta_i Y_{t-i}$$

Then we may be willing to use the same predictor, translated,

$$\hat{Y}_{t+k}(1) = \alpha + \sum_{i=0}^{\infty} \beta_i Y_{t+k-i}$$

at other times. We would then often consider the sequence of one-step-ahead forecast errors, $\varepsilon_t = Y_t - \hat{Y}_{t-1}(1)$, to be exchangeable.

Alternatively, we can think of the observations as being the result of applying a linear filter to a sequence ..., $\varepsilon_{t-2}$, $\varepsilon_{t-1}$, $\varepsilon_t$, of exchangeable random quantities with zero expectation (with additive unknown constants, if required). If we knew the values of $\varepsilon_i$ for $i = t-1$, $t-2$, ... (and any necessary constants), we could calculate $\varepsilon_t$ once $Y_t$ became known and so on.

These ideas are discussed, from a slightly different viewpoint, by Box and Jenkins (1970). The approach recommended by Box and Jenkins involves suitable differencing of non-stationary series until the resulting series may be considered stationary and this has become a standard method. For our purpose, second order stationarity is appropriate and this is also standard, although normality, which makes second order stationarity equivalent to strict stationarity, is also commonly assumed in time-series literature. (Briefly, a sequence $\{Y_t\}$ is second order stationary if the mean of $Y_t$ and the covariance of $Y_t$ and $Y_{t+k}$ do not depend on $t$, for $k = 0, \pm 1, \pm 2, \ldots$ .) For example, differencing once would amount to saying that we expected the mean rate of increase of the series over long periods to remain the same.

Having obtained a stationary series, we can describe its covariance structure by specifying the weights of the (possibly infinite) moving average applied to the exchangeable sequence $\{\varepsilon_t\}$. In practice, we would normally use a finite moving average and/or introduce autoregressive terms which relate $Y_t$ directly to $Y_{t-1}$, $Y_{t-2}$, ... as above (see Box and Jenkins, 1970). Overall, the resulting structure is described as an autoregressive integrated moving average (ARIMA) model. In many practical applications, it will be reasonable to use such a linear model as a description of our beliefs about the observations or some transformation of them.

Autoregressions fit naturally into our modelling framework simply by including nodes for lagged variables. For example, a first order autoregression for the sales, $S_t$, of a product in time period $t$ is represented as in Fig. 4.11(a). Fig. 4.11(b) shows a second order autoregression.

It may appear that moving average processes do not fit so well into our scheme. However, all we need to do is add common uncertainty nodes for the (unobserved) random inputs, $a_t$, say. As an illustration, Fig. 4.11(c) shows a second order moving average process. Note that here $S_t$ is a deterministic node, having zero specific variance given its direct predecessors. In this and most of the subsequent diagrams, we omit the distinction between chance and deterministic nodes and use single circles for both.

Seasonal effects can be incorporated in a variety of ways. An autoregressive possibility is shown in Fig. 4.11(d). Another possibility is to incorporate (unobserved) nodes representing the 'underlying level' for each period of a particular year. This allows prior beliefs about seasonal effects to be used. For other years, a trend, which might be autocorrelated itself, short term fluctuations etc., could be added.

We now apply these ideas to develop a description of beliefs about the time series of sales of beer from the depots. By $V_t = (V_{1t}, \ldots, V_{4t})'$ we represent the volumes of four beers sold by the depots in week $t$. It is, of course, possible to consider individual depots separately. For example depots near the coast may sell increased amounts of some beers in warm summer weather while there may actually be a decrease in city areas. We do

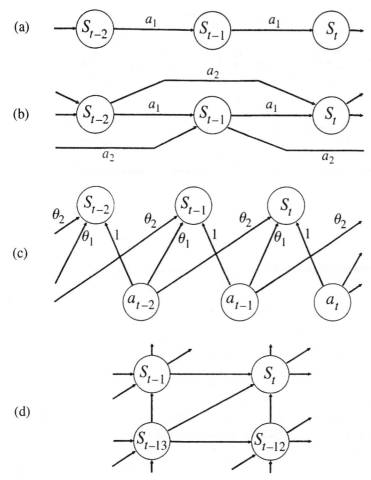

**Fig. 4.11** Time series influence diagrams

not, in this study, consider transformations of the data, for example to render multiplicative seasonal and random components additive. Because of the linkage with stocks, production etc., it is convenient to work directly in terms of volume and, since we are largely interested in forecasts over only a few weeks, this seems unlikely to cause serious problems. In the longer term, changes to scale-dependent quantities such as variances and seasonal effects can be made if required.

When asked how they would forecast, brewery staff said they would look at the sales of the last few weeks and at the same time in the previous year. A straightforward approach in the Box–Jenkins style is not in conflict with this practice and supports our aim of working directly in terms of observables. Other information on future sales may then be used to modify this basic model. Differencing once seasonally and non-seasonally might reasonably be expected to remove seasonal effects and a linear trend. Stationarity with zero mean for the differenced series would suggest a long term forecast function projecting trends linearly. Analysis of past data suggests that, indeed, in the notation of Box and Jenkins (1970), a 'multiplicative' $(0, 1, 1) \times (0, 1, 1)_{52}$ model would be appropriate, i.e.

$$\nabla \nabla_{52} V_{bt} = (1 - \theta_{b1})(1 - \theta_{b52}) \varepsilon_{bt}$$

where $\varepsilon_{bt}$ has zero mean and is uncorrelated with $\varepsilon_{bs}$ for $s \neq t$ and $\nabla = 1 - B$ and $\nabla_{52} = 1 - B^{52}$, where $B$ is the backshift operator such that $BV_t = V_{t-1}$.

Writing $\theta_1$, $\theta_{52}$ for the diagonal matrices with elements given by $\theta_{11}, \ldots,$ $\theta_{41}$ and $\theta_{152}, \ldots, \theta_{452}$ and $\varepsilon_t = (\varepsilon_{1t}, \ldots, \varepsilon_{4t})'$, we can write

$$V_t = V_{t-1}^{(1)} + \varepsilon_t \tag{4.6}$$

where

$$V_{t-1}^{(1)} = V_{t-1} + V_{t-52} - V_{t-53} - \theta_1 \varepsilon_{t-1} - \theta_{52} \varepsilon_{t-52} + \theta_1 \theta_{52} \varepsilon_{t-53} \tag{4.7}$$

is the one-step-ahead forecast.

By observing that $\varepsilon_t = V_t - V_{t-1}^{(1)}$ and substituting, we obtain

$$V_{t-1}^{(1)} = (I - \theta_1)V_{t-1} + \theta_1 V_{t-2}^{(1)} + (I - \theta_{52})V_{t-52}$$
$$+ \theta_{52} V_{t-53}^{(1)} - (I - \theta_1 \theta_{52})V_{t-53} - \theta_1 \theta_{52} V_{t-54}^{(1)}$$

where $I$ is an identity matrix. This corresponds to deleting the $\varepsilon$ nodes in the influence diagram. The resulting diagram is shown in Fig. 4.12. Equation (4.6) shows that all of the information useful for linear prediction of $V_t$, available at time $t$, is contained in $V_{t-1}^{(1)}$.

It is, of course, true that there are not exactly 52 weeks in a year. However, the slow migration out of phase of the model is unlikely to be a problem over the forecast intervals of interest. While we have allowed for the regular seasonal pattern, there are, of course, public holidays and other events with important effects, such as major sporting events, advertising campaigns, etc., which do not recur at the same time each year. Even Christmas, which always occurs on December 25th, falls on a different day of the week each year and this affects the pattern of the holiday period. These features can be dealt with by adding extra nodes to the diagram. For example, in the case of a cricket test match we might add a node representing a temporary extra additive contribution to sales. This could be related to the last occurrence of a similar test match or to other comparable events rather than to 52 weeks ago.

If we had an infinite series of past data and no extra uncertainty from holidays or special events, then we could observe the values of all quantities in Fig. 4.12 since the one-step-ahead forecast $V_{t-1}^{(1)}$ is simply a function of past data and $V^{(1)}$ values. In practice, all $V^{(1)}$ values will have non-zero associated variances.

Fig. 4.13 shows how forecasts are constructed (in the absence of special events). The one-step, two-step, etc. forecasts at time $t$, i.e. $V_t^{(1)}$, $V_t^{(2)}$, $\ldots$, are formed using eq. (4.7). We use eq. (4.6) to calculate $\varepsilon_t$, $\varepsilon_{t-1}$, $\varepsilon_{t-2}$, $\ldots$. Future values $V_{t+1}$, $V_{t+2}$, $\ldots$ are replaced by their expectations given by

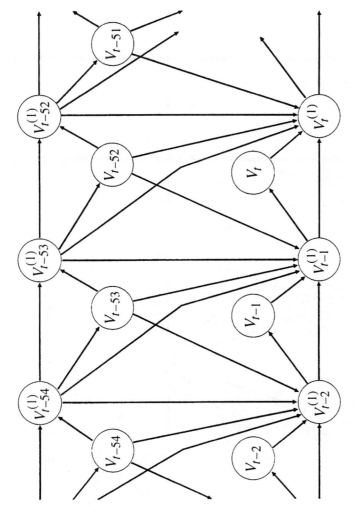

**Fig 4.12** Influence diagram for sales and forecasts

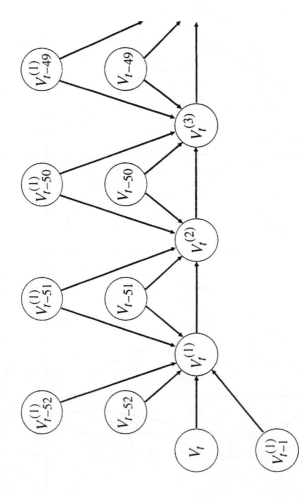

**Fig 4.13** Influence diagram for constructing forecasts

setting $\varepsilon_{t+1}$, $\varepsilon_{t+2}$, ... to their expectation, i.e. zero, in eq. (4.6). Thus:

$$V_t^{(1)} = V_t + V_{t-51} - V_{t-52} - \theta_1 \varepsilon_t - \theta_{52} \varepsilon_{t-51} + \theta_1 \theta_{52} \varepsilon_{t-52}$$
$$V_t^{(2)} = V_t^{(1)} + V_{t-50} - V_{t-51} - \theta_{52} \varepsilon_{t-50} + \theta_1 \theta_{52} \varepsilon_{t-51}$$

etc. Using the substitution for $\varepsilon_t$ etc.,

$$V_t^{(1)} = (I - \theta_1)V_t + \theta_1 V_{t-1}^{(1)} + (I - \theta_{52})V_{t-51}$$
$$+ \theta_{52} V_{t-52}^{(1)} - (I - \theta_1 \theta_{52})V_{t-52} - \theta_1 \theta_{52} V_{t-53}^{(1)}$$
$$V_t^{(2)} = V_t^{(1)} + (I - \theta_{52})V_{t-50} + \theta_{52} V_{t-51}^{(1)} - (I - \theta_1 \theta_{52})V_{t-51} - \theta_1 \theta_{52} V_{t-52}^{(1)}$$

This allows us to predict the values of forecasts which will be made at future times. This can be useful in general prediction where decision rules using the forecasts may be involved. The forecast itself can be regarded as a decision with a linear rule.

To fit Fig. 4.8 into Fig. 4.12, we could, if we wished, add a parent node $\varepsilon_t$ for $V_t$ and a parent node $\eta_t$ for $\varepsilon_t$, with $\varepsilon_t = M\eta_t$, where $M$ is a matrix of appropriate coefficients, as in Fig. 4.14. Node $\varepsilon_t$, which contains the vector $(\varepsilon_{1t}, \varepsilon_{2t}, \varepsilon_{3t}, \varepsilon_{4t})'$, becomes deterministic. Node $V_t$ becomes deterministically equal to $V_{t-1}^{(1)} + \varepsilon_t$. Node $\eta_t$ contains the specific uncertainty factors for 'Total', 'Lager', '$Q$' and $\varepsilon_{3t}$.

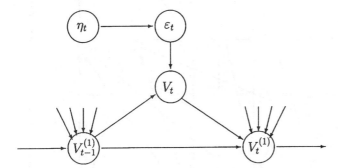

**Fig. 4.14**  Random input to sales time series

While this representation has the merit of being based on observable quantities, a representation in the form of a dynamic linear model may be felt to be more helpful for careful modelling of beliefs. The same covariance structure can, of course, be built up as either an ARIMA model or a DLM, but in the 'state-space' approach we may have several underlying unobserved quantities representing such things as 'trend', and modelling is more directly in terms of these quantities and their variances. This would lead to a simpler diagram and would also make it easier to see how we might add in our beliefs about the effects of holidays and special events and how we might refine the model, for example by allowing neighbouring seasonals to be correlated. We could also allow the relationships between the four beers to be different in the trend, seasonal and 'random' components. For

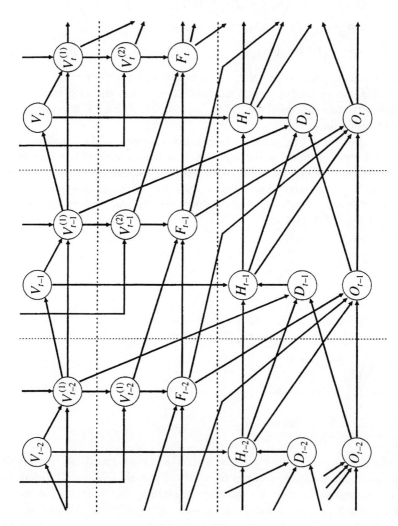

**Fig 4.15** Influence diagram for stocks, orders, sales and deliveries

example, consumption of bitter might be less prone to seasonal or short-term fluctuation than that of lager. On the other hand, we would be introducing quantities which are not observable and the relationship with the forecast is less plain.

Aspects of the representation of dynamic linear models in influence diagrams are discussed by Smith (1990).

### 4.3.5 Linking sales to the planning diagram

Fig. 4.15 is a diagram showing how sales are connected with depot stocks, orders and deliveries on the weekly basis used at the brewery. (For simplicity here, we use the totals over all depots). The nodes are as follows:

- $H_t$ is the volume of stock, above a fixed target level, held in the depots at the end of week $t$.

- $D_t$ is the total deliveries received from the brewery during week $t$. It can be linked directly from the five daily delivery nodes in Fig. 4.2.

- $O_t$ is the total orders, made by the depots in week $t$ for week $t + 1$.

- $F_t$ is the latest available forecast of demand supplied by the depots, namely the depot demand forecast for week $t + 2$ made in week $t$, which is the time when the planning for week $t + 2$ is done.

Those who control deliveries take account of the orders but also of the brewery's own forecasts and the latest available depot stock figures. To predict what the delivery will be we can take into account the order $O_{t-1}$ made in week $t - 1$ for week $t$ and our own prediction of what will be needed to maintain depot stock.

To help us predict the orders, we have available previous order values and also the forecasts supplied by the depots themselves. The latest available depot forecast for week $t + 1$ at the time when the order for week $t + 1$ is made is that made in week $t - 1$, i.e. $F_{t-1}$. We can also use the difference between the depot forecast for week $t$, made in week $t - 2$, and the order actually made for week $t$.

Clearly then we need to be able to predict the depot forecasts. It seems reasonable to believe that the depot forecasts will be related to our own brewery forecasts, but we might expect differences between the forecasts at previous time points to carry over to some extent.

As with the sales themselves, we can adapt our prediction of depot forecasts to allow for holidays and special events. We could also modify our prediction of orders to take account of orders at the same time in the previous year and the pattern of orders around holidays and special events.

The major sources of uncertainty in this diagram are the sales and the depot forecasts, which themselves depend on past sales. We can build our time series model for sales in the upper section of the diagram. The depot forecasts in the central section depend on the sales and finally the uncertainties flow down to the lower section containing the stocks, orders and deliveries.

## 4.4    Elicitation

### 4.4.1    General comment

Once the qualitative structure of a diagram has been built, it is necessary to give the numerical values which allow the means, variances and covariances to be calculated. Although deterministic relationships are sometimes used, dependencies are predictive rather than causal. Uncertainty may lead to the rate of change in the expectation of a quantity, with respect to another on which it depends, being less than would be expected in a deterministic causal relationship. At an early stage in the project, discussion with a manager about the relationship between planned production and actual production led to careful consideration of how such predictive relationships should be elicited.

### 4.4.2    Elicitation of means

In Foresight, unconditional, i.e. marginal, means for nodes are specified directly. However, as explained in Section 4.3.2, there is often a natural ordering of the quantities and we might tend to think of adjusted means given 'earlier' nodes. Therefore it is usually best to work through the diagram, in the direction of the arcs, specifying means and arc coefficients as we go.

The interface for specifying means is shown in Fig. 4.16. Notice that the marginal mean is being requested but the sum of the means of the parent nodes, multiplied by the arc coefficients, is given automatically as a guide.

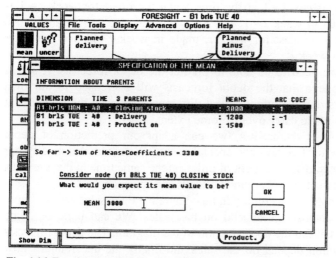

**Fig. 4.16** Foresight interface for specifying means

### 4.4.3   Elicitation of variances

We may elicit marginal variances in Foresight. However, if the values for parent nodes and the incoming arc coefficients are already specified, it is easy to determine the contribution to the marginal variance from the parents and hence work in terms of the specific variance. In these circumstances, Foresight will give the user a lower limit which corresponds to the marginal variance with a zero specific variance. (This constraint is necessary for coherence.) As can be seen in Fig. 4.17, Foresight actually elicits standard deviations rather than variances since the former were found to be more readily understood by users.

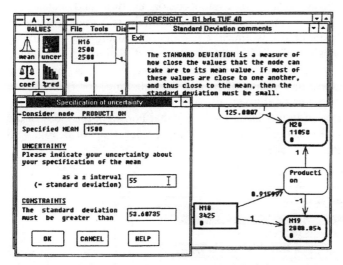

**Fig. 4.17**  Foresight interface for eliciting standard deviations

The variance of a deterministic node is calculated automatically as soon as the variances and covariances of its parent nodes become known and the coefficients of the in coming arcs have been specified.

### 4.4.4   Elicitation of arc coefficients

Arc coefficients are specified directly or indirectly depending on the situation. For example, considering Fig. 4.2, Tuesday's closing stock is equal to Monday's closing stock plus Tuesday's production minus Tuesday's deliveries. Therefore the coefficients of the arcs entering node $BS_{TU}$ are $\pm1$ and can be specified directly.

Even when the coefficient is not as simple as $\pm1$, users may feel able to specify arc coefficients directly. However, this is not always straightforward as there may be a danger of mistaking the value of an arc coefficient for a change in the expected value of a node resulting from a unit change in the value of the corresponding parent node. These are not, in general, the same, since if $A$ is a parent of $B$, learning the value of $A$ may cause a change in the expectations of other nodes which in turn affect the expectation of $B$, as well as the direct effect on $B$. Therefore Foresight provides an alternative elicitation method based on the reduction in variance of a node when a parent node becomes known. Suppose, first, that node $B$ has only one direct predecessor, $A$. Then the reduction in the variance of $B$ when $A$ becomes known is equal to the covariance of $A$ and $B$ which is simply $k_{AB}^2 V_A$, where $k_{AB}$ is the arc coefficient and $V_A$ is the variance of $A$. If $B$ has more than one direct predecessor then the situation is more complicated because of possible covariances among these parent nodes. However, if these covariances among the parents are already elicited then we can elicit the reduction in variance of $B$ given each direct predecessor in turn and hence the covariances with $B$. Determination of the arc coefficients is then a matter of solving a set of simultaneous linear equations and using the previously elicited signs. Foresight also offers the alternative of eliciting the reduction in variance of $B$ given all of its direct predecessors, $A_1$, ..., $A_n$ and hence determining the specific variance of $B$, and then eliciting, for $i = 1$, ..., $n$, the reduction in variance of $B$ given all of the direct predecessors except $A_i$. The only complication in this latter method is that Foresight must adjust the variance of $A_i$ to take account of the other direct predecessors of $B$. Fig. 4.18 shows the Foresight interface.

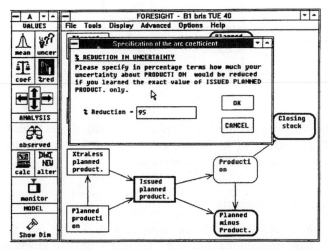

**Fig. 4.18** Foresight interface for eliciting the percentage reduction in variance of a node given the exact value of one of its parents

In the case of decision nodes, such as $XL_{WE}$ in Fig. 4.2, unless a linear rule is used for deciding the value of a decision node, the coefficients of the informational arcs are zero to avoid the danger of inappropriate adjustment of 'past' nodes when a value is given to a decision. Users can, of course, give values to decision nodes to investigate the effects of possible plans.

However, when the value of a decision node is determined by a linear rule then it can contain information about the past.

## 4.5 Managers' view of the system

### 4.5.1 General comments

The computer system was developed with the needs of managers at the Romford Brewery specifically in mind. During the first phase of the project, interviews were conducted with 39 managers from the production, distribution, accounts and engineering departments. An initial appraisal of the information and decision support needs of 24 of these managers was then carried out.

Although the detailed examples used later in the project all referred to the packaging plant, the original intention was that the system could be used throughout the company. Managers in other areas of the company identified uses where the potential advantages of the system were seen. One particularly interesting possibility involved quality control, which could perhaps be treated alongside production since, for example, the rejection of a batch would affect production planning, and decisions about raw materials might affect the quality of the product.

The models constructed using Foresight can be used to provide extensive decision support for management in producing weekly plans for the products, finding solutions which stay within budget, building up stocks in preparation for public holidays, etc. As well as making predictions for what will happen given particular plans, Foresight can also calculate values of decision nodes to make the expectations of other nodes equal to target values.

Thus Foresight can be used:

- to make straightforward predictions of future behaviour;
- to examine the likely effects of scenarios or decisions, i.e. 'what if' analysis;
- to suggest values for decisions.

'What if' analysis may consist of examining the predicted effect of decisions, e.g. production plans, by entering values in decision nodes. Because the arc coefficients entering decision nodes are set at zero, except in cases where an 'automatic' decision by a linear rule is specified, no information is propagated backwards through the diagram. Another kind of 'What if' analysis involves examining the predicted consequences of some circumstance outside of management's control. For example, 'what if sales next month are 50 per cent greater than normal?' Here we can simply enter the values for the particular scenario of interest and see what would be expected to happen to other quantities.

Foresight gives the facility to specify target values for future nodes and to determine, where possible, values for decisions to make the expectation of these nodes equal to the targets (see Spiropoulos, 1995, Appendix 17). In this way suggestions are given for the values of decisions.

### 4.5.2   Usability study

In order to obtain an indication of how easy or difficult industrial managers find using Foresight, eight 'usability' sessions were organized in spring 1994. The users were eight male managers from Carlsberg-Tetley Brewing Limited. Following the distribution of a memorandum, the users volunteered for the study, and sessions were held at three different locations in the UK. Each session lasted between 1 and 1.5 hours, and each user filled in a questionnaire which consisted of 38 questions. A full description of the usability study with discussion is included in Spiropoulos (1995). When planning the sessions, it was thought that if Foresight crashed or did not perform as expected, or if the user did not like a particular aspect of the interface, this could influence the user's responses to the questions designed to investigate the elicitation methods. Therefore, the users were made aware that the purpose of the experiments was to address the following three issues, and that each issue should be considered separately:

- How easy is it to use the interface of Foresight to create the qualitative and quantitative structure of BLIDs?

- How easy is it to use a fully specified model in Foresight for decision support purposes?

- How easy is it to elicit means, standard deviations, arc coefficients, coefficients of percentage reductions in uncertainty and adjusted expectations using the developed elicitation methods?

Furthermore, when the questionnaire was designed it was decided to obtain an indication of how familiar the users were with means, standard deviations, variances, covariances and correlation coefficients, and how often they used these, and to include questions about some general issues, such as how useful is it to see and explore the internal structure of models in the form of a BLID.

During the first 10 minutes the users were given a brief introduction to BLIDs and an initial 'familiarization' with Foresight's interface. Then, the users were given an extract of the BLID shown in Fig. 4.2 containing days Monday and Tuesday only, and were asked to draw it in Foresight. That subdiagram was chosen because it could be explained quickly and involved quantities such as production, delivery and closing stock with which managers were already familiar. The full names of the nodes were displayed on the diagram as shown in Fig. 4.19. The users were asked to perform the following tasks in the following order:

- Draw the part of the diagram that appears on the left of the dotted line in one dimension, and name the dimension 'Monday' (namely, develop Fig. 4.19).

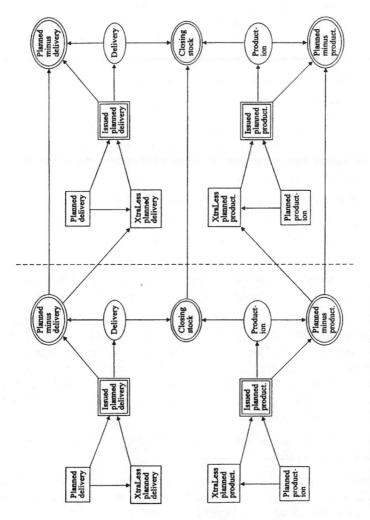

**Fig 4.19** BLID representing Monday's qualitative structure

- Draw the part of the diagram that appears on the right of the dotted line in a dimension called 'Tuesday' by copying the qualitative structure of 'Monday'.

- Draw the arcs that cross the dotted line in the diagram as cross-dimension arcs between 'Monday' and 'Tuesday'.

- Insert values into the nodes and arcs of 'Monday', and specify the arc coefficients between issued planned production and production, and issued planned delivery and delivery, indirectly using the percentage reduction in uncertainty method (icon *%red*).

- Specify the coefficients of the cross-dimension arcs.

- Load a fully specified 3-day version of the diagram.

- Use Foresight for decision support purposes and read the generated reports. Formulate 'What if' analyses queries and initiate the adjustment, and set targets and initiate the calculation for identifying matching inputs for the chosen decision nodes.

The results of the questionnaire can be summarized as follows:

- The users had a varying degree of familiarity and usage of graphical interfaces and statistical quantities, ranging from 'very familiar' to 'very unfamiliar'.

- All users found Foresight's interface for developing BLIDs and using fully specified models for decision support purposes user-friendly.

- The majority of the users could specify means, standard deviations and arc coefficients directly without any problems, and only one user had real difficulties. Regarding the two methods used for the elicitation of arc coefficients and covariances indirectly by specifying the reductions in variance given one or all but one parent, all users indicated that if both elicitation methods were available they would be able to respond to one of them without any difficulty.

- All users thought that it is useful to see and explore the BLIDs that form the internal structure of the models.

- All users recognized the importance of providing summaries of information flow and diagnostics to check and clarify the prior specification of beliefs and the extent of the belief adjustment.

## 4.6   Diagnostics

### 4.6.1   Introduction

Any model-building process consists of iterative cycles of model modification and diagnostic checking or validation. The process of developing a Bayes-linear predictive belief specification, for example with Foresight, follows this cycle. Before any data are observed, the belief specification process can be assisted by tools which help the user to examine and check features of the developing specification. After data are observed, the user's attention can be drawn to cases where the data appear to be unusually out of line with prior expectations in some way.

### 4.6.2   Examining the belief structure

Before any data are observed we can quantify the influence that one node exerts on another and show the extent to which the uncertainty of a node is reduced if others are observed.

An obvious and simple feature provided by Foresight is that, if a change in a belief specification leads to incoherence, this is immediately detected and a warning messages is given to the user. This can happen, for example, simply as a result of changing the variance of a node. In Fig. 4.20 a change in the variance of node N1 has caused the specific variances of nodes N3 and N4 to become negative, which indicates incoherence. Foresight gives a warning message and indicates the problem nodes. We can see in the display shown the negative values of *%rem*, the 'remaining' variance, when the parent nodes become known, as a percentage of the marginal variance. In a similar way *%red* is the reduction in variance as a percentage of the marginal variance. The figures shown on the arcs in this display represent the percentage reductions in variance given only the particular parent node concerned. Other displays available in Foresight include the actual marginal and specific variances, the arc coefficients and the parent–child covariances.

These displays, currently available in Foresight, are a simplified version of the displays discussed in Goldstein *et al.* (1993) which are more fully

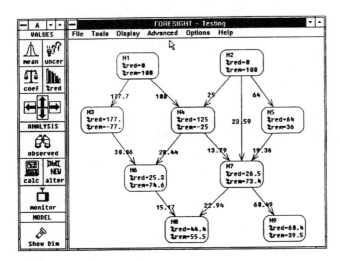

**Fig. 4.20** Flow of information summaries

implemented in another computer program, [B/D] (see, e.g., Goldstein and Wooff, 1995, for a description of [B/D]). It is hoped, when time permits, to upgrade the diagnostic capabilities of Foresight and perhaps to combine the greater analytic capabilities of [B/D] with a Foresight-style graphical interface.

### 4.6.3 Diagnostics after observing data

When data have been observed, for example actual beer sales, production etc., we can use diagnostics to compare the changes in expectation of various quantities with the sizes of changes which were expected. When a quantity is itself observed, this is just a special case of a change in expectation, being the final change. We can also show which arcs in the diagram are 'responsible' for any surprising changes. Graphical displays for these purposes are described in Goldstein *et al.* (1993). These are not yet implemented in Foresight and the display in Fig. 4.21 was constructed in [B/D] (see Goldstein and Wooff, 1995). This corresponds to Fig. 4.15 extended over 8 consecutive weeks. Apart from the first week, each node here has an inner and an outer shading. The outer shading represents the division of the uncertainty in the node into that part which is resolved when adjustment is made by parent nodes and that part which remains until the node itself is observed. Corresponding to each of these outer sectors, the inner sectors are shaded to indicate the standardized difference between the size of the actual change in expectation, when the corresponding

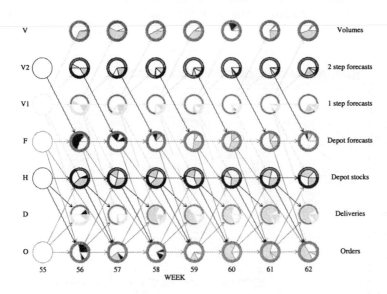

**Fig. 4.21** Diagnostic display

observations are made, and the expectation for this change. The amount of shading represents the size of the difference and the type, dark or light, or red and blue on a colour screen, whether the change was greater or less than expected. The scaling is such that shading of about half of an arc signals a diagnostic warning that changes are substantially larger or smaller than expected. In Fig. 4.21, with a few exceptions, the shading is light, suggesting that we have built too much variation into the system and that our short-term predictions are mostly more accurate than we had supposed.

## 4.7 Conclusions

In this account, we have described the elements of the construction of a Bayes-linear decision support system for managers in a brewery. The Bayes-linear approach is similar in spirit to a full Bayes analysis; for example, many of the calculations are the same as for a full Bayes analysis assuming multivariate normality. However, in our approach, we do not make any distributional assumptions. Rather, if all that the manager is able to specify is the mean and variance structure for the problem, then the Bayes-linear approach makes sensible use of this restricted prior specification to produce an integrated graphical system to support forecasting and decision making. The essential stages in building this system are the construction of a qualitative representation of the structure of the beliefs of the manager, using a graphical interface such as Foresight, followed by a careful quantification of means and variances over the graphical structure. As we receive information, we may then subject the forecasts of the system to careful diagnostic scrutiny so that we are able to judge and improve the reliability of the system over time. In our experience, this is a natural and simple approach for treating decisions and uncertainties in many complex systems, and one which managers may readily appreciate and apply.

## References

Box, G.E.P. and Jenkins, G.M. 1970: *Time series analysis: forecasting and control.* San Francisco: Holden-Day.

Farrow, M. and Goldstein, M. 1992: Reconciling costs and benefits in experimental design. In Bernardo, J.M., Berger, J.O., Dawid, A.P. and Smith, A.F.M. (eds) *Bayesian statistics 4.* Oxford: Oxford University Press, 607–15.

Farrow, M. and Goldstein, M. 1993: Bayes linear methods for grouped multivariate repeated measurement studies with application to crossover trials. *Biometrika* **80**, 39–59.

Farrow, M. and Spiropoulos, T. 1992: Graphical partial belief representation and linear Bayes prediction for management, *Occasional Paper 92–9.* University of Sunderland: School of Computing and Information Systems.

Goldstein, M. 1990: Influence and belief adjustment. In Oliver, R.M. and Smith, J.Q. (eds) *Influence diagrams, belief nets and decision analysis.* Chichester: Wiley, 143–74.

Goldstein, M., Farrow, M. and Spiropoulos, T. 1993: Prediction under the influence: Bayes linear influence diagrams for prediction in a large brewery. *The Statistician* **42**, 445–59.

Goldstein, M. and Wooff, D. 1995: Bayes linear computation: concepts, implementation and programs. *Statistics and Computing* **5**, 327–41.

Smith, J.Q. 1989: Influence diagrams for statistical modelling. *Annals of Statistics* **17**(2), 654–72.

Smith, J.Q. 1990: Statistical principles on graphs. In Oliver, R.M. and Smith, J.Q. (eds), *Influence diagrams, belief nets and decision analysis*. Chichester: Wiley, 89–120.

Spiropoulos, T. 1995: Decision support for management using Bayes linear influence diagrams. *PhD thesis*, University of Sunderland.

# 5 Bayesian methods in reservoir operations: the Zambezi river case[1]

D Ríos Insua, K A Salewicz, P Müller, C Bielza

## 5.1 Reservoir operations

Many reservoirs have been built and are operating under different climatic and hydrological conditions, and with many different purposes, including municipal, industrial or agricultural water supply, hydropower generation, flood protection and control and recreation. In fact, many reservoirs serve several conflicting purposes, as in the case of having to secure water supply to a number of users and to provide flood protection downstream.

Multiple objectives are not the only difficulties associated with reservoir operations. The inflow process is typically uncertain and has varying dynamics. Water demand and priorities may also change from one period to another. Consequences of operational decisions might affect different groups of people in different ways. These and many other features of reservoir operation problems mean that, despite a long tradition of efforts aimed at developing efficient and secure methods for reservoir operation, no generally accepted methodology is available as yet.

In this paper, we provide a general Bayesian methodology and describe its application to the main reservoirs in the Zambezi river.

### 5.1.1 Quantitative methods in reservoir operation

The literature dealing with various aspects of single and/or multiple reservoir management is rich and covers a broad range of methods. Since there is no single, general formulation for the problem, there is no universally applicable solution approach. We shall concentrate on reservoir operation problems, leaving aside related problems of reservoir design (site and size). Thus, our problem is as follows: given an existing reservoir, find the operating policy or set of rules specifying how much water has to be released from the reservoir for different purposes and at various times of interest.

[1] This research was partially supported by an Iberdrola Foundation project and grants from CICYT and NATO. It was completed while David Ríos Insua visited CNR-IAMI.

The very first methods in the field developed in the 19th century were based on simple diagrams (Klemes, 1981). Since then, both single and, much less frequently, multiple objective optimization methods have been proposed. Also, uncertainty has been included, both explicitly (via probabilistic models and techniques) and implicitly (via scenarios). We believe, though, that the literature dealing with reservoir operation problems typically identifies operations with a specific technique applied to solve a certain formulation of the problem, leading to the false conclusion that the reservoir operation task can be solved by direct application of the optimization method described in a particular paper. This 'technique-focused' approach fails to present the comprehensive framework necessary to tackle reservoir operation problems without too many simplifying assumptions. The most common deficiency of various 'technique-focused' approaches boils down to a lack of mechanisms for using incoming information about inflow and users' demands and preferences, thus limiting the use of up-to-date information concerning reservoir operation conditions.

### 5.1.2   Bayesian methods in reservoir operation

We aim to solve the problem within the Bayesian framework. This requires:

- defining the set of control alternatives, in our case release policies $u$;
- providing a forecasting model for inflows $i$ to the reservoir;
- modelling the consequences $c(u, i)$, associated with policy $u$ and inflow $i$;
- modelling the preferences of the decision maker by means of a utility function $F$, typically multi-objective and multiperiod.

Let $h$ denote the predictive density of inflows and $D$ the inflow history. Let the subscript $j$ refer to period $j$, let $s_{t+k+1}$ be the final storage, and let $k$ be the planning horizon. Then, at time $t$, the reservoir management planning problem consists of finding the controls $(u_t, ..., u_{t+k})$, maximizing the expected utility

$$\int F\left[c(u_t, i_t), ..., c(u_{t+k}, i_{t+k}), s_{t+k+1}\right] h(i_t, ..., i_{t+k} | D_t)\, di_t ... di_{t+k}$$

taking into account the dynamics of the reservoir system and the constraints over controls and reservoir storages. The problem becomes unmanageable for a realistic planning horizon, say 36 months, since we have to solve a long-term stochastic dynamic program, the evaluation of each control requires the solution of a high-dimensional integral, and uncertainty about the inflow process rapidly propagates through time.

Instead, we adopt a strategy based on a concept of 'reference trajectory', which assumes having found a good 'reference' storage level for each period. Then, at each period we would like to maximize the expected value of a utility function $F^*$ taking into account the consequences of interest and the deviation from that reference state, i.e.

$$\int F^* \left[ c(u_t, i_t), \delta(s_{t+1}, s^*_{t+1}) \right] h(i_t | D_t) \, di_t$$

where $\delta(s_{t+1}, s^*_{t+1})$ represents the deviation of the final state $s_{t+1}$ from the reference state $s^*_{t+1}$. Intuitively, if the reference states are defined in such a way as to account for the dynamic aspects of the problem, we would not lose too much with this approach.

This proposal is, in fact, based on a traditional and well established method of reservoir operation using the concept of *rule curves*, which represent optimal trajectories of the reservoir over a long time horizon (see Loucks and Sigvaldasson, 1982).

## 5.2   The Zambezi river case study

The Zambezi is the largest African river flowing into the Indian Ocean. Its total catchment area covers 1300 000 km², while its length from the source in the Central African Plateau to the outlet is in the order of 2500 km (Balek, 1977). The catchment is shared by eight riparian countries; for Mozambique, Zambia and Zimbabwe it constitutes the main water resource. The main use of the river is hydropower generation. The main reservoirs are Lake Kariba and Cahora Bassa (Fig. 5.1).

Lake Kariba, on the border between Zambia and Zimbabwe, is currently the fourth largest man-made lake in the world. At the maximum retention level, it covers an area of over 5600 km² and has an active storage exceeding 70 km³. Hydropower plants installed at the northern (Zambian) and southern (Zimbabwean) banks of the dam plus a small hydropower scheme located on the Kafue River jointly supply more than 70 per cent of the electricity

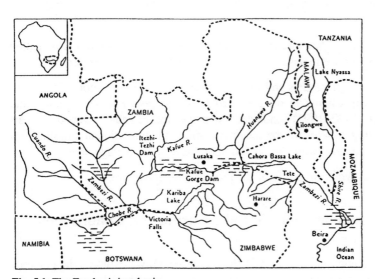

**Fig. 5.1** The Zambezi river basin

produced in these two countries. Since the completion of the generating facilities in 1977, the Lake Kariba system has supplied a monthly average of about 600 GWh/month, with little seasonal variation. Both countries operate the scheme jointly and share the electricity generated on a 50 : 50 basis.

Cahora Bassa is located in Mozambique. The maximum area is 2700 km$^2$. The maximum impounded water is 60 km$^3$. The energy produced is mainly sold to South Africa. However, until recently, the transmission lines have been a favourite terrorist target, limiting its operation.

### 5.2.1   The IIASA project

Upon request from the governments of Botswana, Zambia and Zimbabwe, the United Nations Environment Program (UNEP) assisted the governments of the Zambezi basin countries in developing the Zambezi action plan (ZACPLAN). In May 1987, an agreement on the 'Action plan for the environmentally sound management of the common Zambezi river system' was adopted and signed in Harare (Zimbabwe) by representatives of five basin countries (UNEP, 1987). ZACPLAN was then approved by the summit of the Southern African Development Coordination Conference and a special unit was established to implement the plan.

At the same time, UNEP recognized the need to improve government decision making processes, in selecting the best environmentally sound alternatives in planning and managing large international water systems. The International Institute for Applied Systems Analysis in Laxenburg (Austria) was then asked to contribute to the work and develop appropriate methods and tools that could be used to support decision making processes associated with the management of Zambezi resources (Salewicz and Loucks, 1988; Salewicz *et al.*, 1989).

IIASA studies in connection with the Zambezi case covered several issues (Pinay, 1988) but focused mainly on two topics:

- development of an interactive river system simulation program, called IRIS, now in use in several countries (see Venema and Schiller, 1995);

- development of models and methods to improve the operation of the main hydropower scheme in the Zambezi basin at Lake Kariba. Gandolfi and Salewicz (1990, 1991) conducted extensive studies and developed two sets of operating policies for the scheme, combining simulation and multiobjective optimization, which improved upon the current management.

These studies were the origin of the project described here.

## 5.3   Bayesian methods in Lake Kariba and Cahora Bassa operation

In Ríos Insua and Salewicz (1995) (RS hereinafter), we explored the potential of the Bayesian framework described above and improved significantly the performance of Lake Kariba. Here we demonstrate the applicability of our framework to multireservoir systems in the cases of Lake Kariba and Cahora Bassa. As explained in Lamond *et al.* (1995), stochastic optimization of multireservoir river basin models is bedevilled by the 'curse of dimensionality'. We suggest our methodology as a powerful alternative.

In the study, one of us (KAS), drawing on experience of the operation of the Lake Kariba over an extended period, plays the role of an expert providing beliefs and preferences, which are then thoroughly checked via sensitivity analysis. This is a common practice in public policy decision analysis (Keeney, 1992).

As far as notation is concerned, superscript 'c' will refer to Cahora Bassa, whereas superscript 'k' will refer to Lake Kariba. For example, $i_t^c$ and $i_t^k$ will designate the inflows (in mln m$^3$/month) to Cahora and Kariba during month $t$, respectively.

### 5.3.1 Operating policies

Let us start by defining our operating policies. Current ones are defined by means of parametric curves associating release with the amount of water stored in the reservoir. The parameters of these curves were chosen to optimize some performance function. Instead, we prefer to formulate more flexible rules, as described for Cahora Bassa:

- At the beginning of the month, the operator announces the desired amounts of water $u_{1t}^c$ and $u_{2t}^c$ (in mln m$^3$/month) to be released, respectively, for energy production and for free storage provision to catch expected floods.

- If there is not enough water to release $u_{1t}^c$, all available water is released for energy production. Otherwise, $u_{1t}^c$ is released for energy production.

- If, after the release of $u_{1t}^c$, there is still water available, some water may be additionally released to control the reservoir storage level. If there is not enough water to release the volume $u_{2t}^c$ announced, all available water is released. Otherwise, $u_{2t}^c$ is released. In the event that, after the two releases, the remaining water would exceed the maximum storage $M^c = 60\ 000$ mln m$^3$, all excess water is spilled.

For Lake Kariba, there are corresponding releases $u_{1t}^k$ and $u_{2t}^k$, defined in a similar vein, with maximum storage $M^k = 70\ 980$ mln m$^3$.

We shall find optimal controls $u_1^c$, $u_2^c$, $u_1^k$, $u_2^k$ associated with our model. Note that actual releases will differ from the announced ones, depending on the available water.

### 5.3.2 Consequences of operating policies

From the operational and managerial viewpoint, the consequences of an operating policy for Cahora Bassa at the end of every month are:

1. the amount of energy produced, $E_t^c$, in GWh/month;

2. the reservoir storage $s_t^c$ in mln m³.

For Kariba the relevant consequences are:

1. the existence of energy deficit $k_t$, with a target of 750 GWh/month;

2. the volume of spilled water $u_{2t}^k$;

3. the volume of water released $u_{1t}^k + u_{2t}^k$;

4. the reservoir storage $s_t^k$.

Note the differences for both reservoirs: first, energy at Cahora Bassa is eventually sold to other countries, whereas energy at Kariba is consumed within the producing countries. The second consequence of interest for Kariba is due to an objective related to homogeneity in operation through time, for reasons described in Gandolfi and Salewicz (1991). The third one relates to the need to take into account the release from Kariba in the inflow model to Cahora Bassa. Reservoir storages are required to keep track of the evolution of reservoirs through time.

An important feature in the problem is the dynamics of the basin. Here we describe Cahora Bassa dynamics. Interreservoir dynamics are described in Sections 5.3.3.2 and 5.3.3.3. Lake Kariba dynamics are described in RS.

Let $u_t^c$ and $e_t^c$ denote, respectively, the amounts of water released and evaporated during month $t$, with $i_t^c$ and $s_t^c$ defined as above. A continuity equation describes the relation between storage level, inflow, outflow and evaporation:

$$s_{t+1}^c = s_t^c + i_t^c - u_t^c - e_t^c$$

Total outflow is

$$u_t^c = u_{1t}^c + u_{2t}^c \tag{5.1}$$

For evaporation, we use a model

$$e_t^c = m_t \left( a \times \frac{s_t^c + s_{t+1}^c}{2} + b \right)$$

where $m_t$ represents evaporation intensity during month $t$. Simple computations lead to a new version of the continuity equation:

$$s_{t+1}^c = d_{1t}^c s_t^c + d_{2t}^c (i_t^c - u_t^c) + d_{3t}^c \tag{5.2}$$

where $d_{1t}^c$, $d_{2t}^c$, $d_{3t}^c$ are appropriate periodic constants obtained from simple transformations and are in Table 5.1. Active storage, in mln m³, should be not less than a minimum volume required to maintain fish population life and not greater than the maximum, i.e.

$$1344 \leqslant s_t^c \leqslant M^c \tag{5.3}$$

**Table 5.1** Evaporation constants for Cahora Bassa

| Month | $d_1^c$ | $d_2^c$ | $d_3^c$ |
|---|---|---|---|
| October | 0.994 | 0.997 | 989.640 |
| November | 0.996 | 0.998 | 990.375 |
| December | 1.000 | 1.000 | 992.395 |
| January | 1.000 | 1.000 | 992.476 |
| February | 1.000 | 1.000 | 992.807 |
| March | 0.998 | 0.999 | 991.691 |
| April | 0.996 | 0.998 | 990.772 |
| May | 0.996 | 0.998 | 990.567 |
| June | 0.996 | 0.998 | 990.737 |
| July | 0.997 | 0.998 | 990.965 |
| August | 0.996 | 0.998 | 990.553 |
| September | 0.995 | 0.997 | 990.016 |

We assume that four generating units are operational (out of five). The maximum water through each turbine is 1212.19 mln m$^3$/month. Therefore

$$0 \leqslant u_{1t}^c \leqslant 4848.76 \qquad (5.4)$$

There are also constraints on the volume $u_{2t}^c$ of spilled water. Given the structure of the dam, release through spillgates will depend on the level of the reservoir. We provide the maximum release as a function of the level $l_t$, measured in metres above sea level:

$$0 \leqslant u_{2t}^c \leqslant \begin{cases} 36761 & \text{if } l_t \geqslant 326 \\ 3423.822\sqrt{l_t - 210} & \text{if } 320 \leqslant l_t < 326 \\ 3338.186\sqrt{l_t - 210} & \text{if } l_t < 320 \end{cases} \qquad (5.5)$$

We provide a relation between storage and level, which we obtain by fitting a least-squares curve:

$$l_t = l_t(s_t^c) = \alpha[1 - \exp(\tau s_t^c)] + \gamma$$

The estimates are

$$\alpha = 48.83$$

$$\tau = -0.000\ 018\ 3393$$

$$\gamma = 296.11$$

Obviously, it is important to provide a relation for the energy produced. We first relate head $h_t$, level $l_t$ and tailrace $r_t$, in metres:

$$h_t = l_t - r_t$$

Using available reservoir design data, we fit a model, by least squares, relating tailrace and release,

$$r_t = \beta \log(u_t^c) + \delta$$

with estimates $\beta = 10.3816$, $\delta = 113.8537$. Finally, the energy produced in GWh/month is

$$E_t^c = \eta u_{1,t}^c h_t \tag{5.6}$$

with

$$\eta = 0.002\ 725 \times 0.85 \times 0.778 = 0.001\ 802\ 04$$

We may then build Table 5.2 showing consequences of various release policies $u_1^c$, $u_2^c$, as a function of the inflow $i^c$ and the available amount of

**Table 5.2** Consequences for Cahora Bassa policies. Case in which $h_1 \geqslant 0$

| Inflow | Final storage | Energy |
|---|---|---|
| $i^c \leqslant h_1$ | 0 | $\eta \dfrac{h(i^c)}{d_2^c} \left[ l(s^c) - \beta \log\left(\dfrac{h(i^c)}{d_2^c}\right) - \delta \right]$ |
| $h_1 < i^c \leqslant h_2$ | 0 | $\eta u_1^c \left[ l(s^c) - \beta \log\left(\dfrac{h(i^c)}{d_2^c}\right) - \delta \right]$ |
| $h_2 < i^c \leqslant h_3$ | $d_2^c i^c - d_2^c h_2$ | $\eta u_1^c [l(s^c) - \beta \log(u^c) - \delta]$ |
| $i^c > h_3$ | $M^c$ | $\eta u_1^c \left[ l(s^c) - \beta \log\left(s^c \dfrac{d_1^c}{d_2^c} + i^c + \dfrac{d_3^c}{d_2^c} - \dfrac{M^c}{d_2^c}\right) - \delta \right]$ |

water in Cahora Bassa $h(i^c) = d_1^c s^c + d_2^c i^c + d_3^c$, where, for simplicity, we drop subscripts $t$. We use the notation:

$$h_1 = \frac{d_2^c u_1^c - d_1^c s^c - d_3^c}{d_2^c}$$

$$h_2 = \frac{d_2^c u^c - d_1^c s^c - d_3^c}{d_2^c}$$

$$h_3 = \frac{M^c + d_2^c u^c - d_1^c s^c - d_3^c}{d_2^c}$$

and consider only the case $0 \leqslant h_1$, in which if $i^c = 0$, we would not be able to meet the operator's demands concerning water through turbines, i.e. $u_1^c \geqslant h(0)/d_2^c$.

Similarly, when the inflow to Kariba is $i^k$ the available water in the

Kariba reservoir will be $g(i^k) = d_1^k s^k + d_2^k i^k + d_3^k$. Then, if

$$g_1 = \frac{d_2^k u_1^k - d_1^k s^k - d_3^k}{d_2^k}$$

$$g_2 = \frac{d_2^k u^k - d_1^k s^k - d_3^k}{d_2^k}$$

$$g_3 = \frac{M^k + d_2^k u^k - d_1^k s^k - d_3^k}{d_2^k}$$

$$g_4 = g_1 + \frac{M^k}{d_2^k}$$

where $d_1^k$, $d_2^k$, $d_3^k$ are evaporation constants for Kariba, we have Table 5.3 as our table of consequences (see RS for further details). Again, we have included only the case $0 \leq g_1$, in which, if there was no inflow to Kariba, we

**Table 5.3** Consequences for Kariba policies. Case in which $g_1 \geq 0$

| Inflow | Spillage | Total release | Final storage | Inflow | Def.* |
|---|---|---|---|---|---|
| $i^k \leq g_1$ | 0 | $g(i^k)/d_2^k$ | 0 | | |
| $g_1 < i^k \leq g_2$ | $i^k - g_1$ | $g(i^k)/d_2^k$ | 0 | $i^k \leq o_1$ | 0 |
| $g_2 < i^k \leq g_3$ | $u_2^k$ | $u_1^k + u_2^k$ | $d_2^k(i^k - g_2)$ | | |
| $i^k > g_3$ | $i^k - g_4$ | $[g(i^k) - M^k]/d_2^k$ | $M^k$ | $i^k > o_1$ | 1 |

*Def. = deficit of energy with respect to the setup target.

would not be able to meet the operator's demands concerning water through turbines. $o_1$ is a minimum inflow level not leading to energy deficit, as a consequence of the energy produced being a non-decreasing function of the inflow.

### 5.3.3  Forecasting models

The main source of uncertainty in reservoir management is associated with the inflow process: monthly releases should take into account the inflow in the corresponding and later months, which are uncertain. In order to solve the forecasting problem, we employ Bayesian dynamic models (DM). West and Harrison (1997) and West (1995) described these models comprehensively.

Given the location of both reservoirs, we consider a forecasting model for the inflows to Kariba, which we only outline (see RS for details), and a forecasting model for the inflows to Cahora Bassa. This one will depend on those inflows not coming from Kariba, which we call incremental inflows, and on the releases from Kariba. This accounts for interreservoir dynamics.

### 5.3.3.1   Inflows to Lake Kariba

After log-transformation, we modelled the Kariba inflow time series with a level term, a term representing seasonal (annual) variation, and a low coefficient, first order autoregressive term to improve short-term forecasts. We only retained the first harmonic in the seasonal part (Fig. 5.2). Consequently, we ended up working with the following dynamic linear model (DLM).

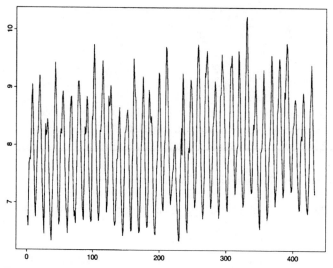

**Fig. 5.2** Logarithm of Kariba inflows (in mln m$^3$)

*Observation equation*

$$y_t^k = z_t^{1k} + z_t^{2k} + z_t^{4k} + v_t^k, \quad v_t^k \sim N(0, v^k)$$

where $y_t^k = \log(i_t^k)$ is the logarithm of the inflow to Kariba; $z_t^{1k}$ designates the level of the series; $z_t^{2k}$ and $z_t^{3k}$ refer to the seasonal term (see below); $z_t^{4k}$ refers to the autoregressive term; and $v_t^k$ designates a Gaussian error term of constant, but unknown, variance $v^k$.

*System equation*

$$z_t^{1k} = z_{t-1}^{1k} + w_t^{1k}$$

$$z_t^{2k} = \cos(\pi/6)z_{t-1}^{2k} + \sin(\pi/6)z_{t-1}^{3k} + w_t^{2k}$$

$$z_t^{3k} = -\sin(\pi/6)z_{t-1}^{2k} + \cos(\pi/6)z_{t-1}^{3k} + w_t^{3k}$$

$$z_t^{4k} = 0.4z_{t-1}^{4k} + w_t^{4k}$$

with $\mathbf{w}_t^k = (w_t^{1k}, w_t^{2k}, w_t^{3k}, w_t^{4k})$ being an error term such that

$$\mathbf{w}_t^k \sim N\left(0, \begin{pmatrix} v^k W^{*k} & 0 \\ 0 & \sigma^{k2} \end{pmatrix}\right)$$

where $\sigma^{k2}$ is the autoregressive variance, and $W_t^{*k}$ is the variance matrix (up to $v^k$) of the first three terms. This matrix was defined using discounting, with a discount factor of 0.8 for the level and a discount factor of 0.95 for the seasonal part.

*Prior information*

$$z_0^k \mid \phi^k \sim N(\mathbf{m}_0^k, v^k C^{*k})$$

$$\phi^k \sim \text{Gamma}(n_0^k/2, d_0^k/2)$$

with $z_0^k = (z_0^{1k}, z_0^{2k}, z_0^{3k}, z_0^{4k})$ and $\phi^k = 1/v^k$. Prior parameters were specified judgementally, with $\mathbf{m}_0^k$ as $(7.8, -1.02, 0.33, 0)$,

$$C^{*k} = \begin{pmatrix} 0.02 & 0 & 0 & 0 \\ 0 & 0.002 & 0.0007 & 0 \\ 0 & 0.0007 & 0.003 & 0 \\ 0 & 0 & 0 & 0.1 \end{pmatrix}$$

$n_0^k = 10$ and $d_0^k = 0.8$, and sensitivity was thoroughly checked.

### 5.3.3.2  *Dynamic regression Kariba–Cahora Bassa*

Exploratory data analyses suggested regressing dynamically the inflows to Kariba and Cahora Bassa, after a log-transformation (*see* Fig. 5.3). For

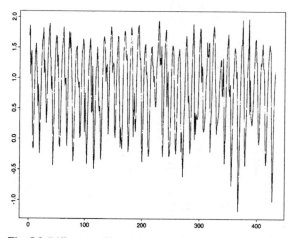

**Fig. 5.3**  Difference of logarithms of inflows to Cahora Bassa and Kariba

this, we define $y_t^c = \log(i_t^c)$ and $z_t = y_t^c - y_t^k$. We try to forecast $z_t$, plotted above, again through a level and a seasonal part. In this case, since the expert had little feeling about this variable, we based the assessments on the data as reflected in Table 5.4. We checked them afterwards via sensitivity analysis.

**Table 5.4** Assessments for level and seasonal effects

|  | Level | Oct | Nov | Dec | Jan | Feb | Mar | Apr | May | Jun | Jul | Aug | Sep |
|---|---|---|---|---|---|---|---|---|---|---|---|---|---|
| Mean | 0.85 | 1.64 | 1.59 | 1.06 | 0.91 | 0.85 | 0.41 | −0.16 | −0.04 | 0.36 | 0.86 | 1.23 | 1.46 |
| Effect |  | −0.79 | −0.74 | −0.21 | −0.06 | −0.00 | 0.43 | 1.01 | 0.89 | 0.48 | −0.01 | −0.38 | −0.61 |
| Maximum deviation | 0.33 | 0.35 | 0.38 | 1.06 | 1.11 | 0.98 | 1.13 | 1.01 | 0.85 | 0.68 | 0.69 | 0.49 | 0.39 |
| Standard deviation | 0.16 | 0.00 | 0.02 | 0.36 | 0.39 | 0.32 | 0.39 | 0.33 | 0.25 | 0.17 | 0.17 | 0.08 | 0.03 |

Variances were deduced accordingly, assuming independence between level and monthly effects. For example, for the level, the maximum deviation from the expected value is about 0.33 and we assume a standard deviation of 0.165.

The Fourier decomposition of the seasonal part is shown in Table 5.5,

**Table 5.5** Fourier decomposition of seasonal effects

| Harmonic | 0 | 1 | 2 | 3 | 4 | 5 | 6 |
|---|---|---|---|---|---|---|---|
| $a_i$ | | −0.779 | 0.056 | −0.122 | 0.033 | −0.003 | 0.017 |
| $b_i$ | − | −0.103 | 0.143 | −0.091 | −0.048 | −0.012 | − |

with covariance matrices deduced from those of the effects. For example, for the first harmonic,

$$C_1^* = \begin{pmatrix} 0.017 & 0.005 \\ 0.005 & 0.015 \end{pmatrix}$$

The relevance of harmonics was assessed with methods described in West and Harrison (1997) with results as shown in Table 5.6. As a consequence,

**Table 5.6** Relevance of harmonics

| Harmonic | 1 | 2 | 3 | 4 | 5 | 6 |
|---|---|---|---|---|---|---|
| Statistic | −5.77 | −0.15 | 0.49 | −0.28 | −0.00 | 0.00 |

only the first harmonic was retained initially. However, posterior analyses suggested that the inclusion of the second harmonic would be beneficial, together with a low coefficient, first order autoregressive term to improve short-term forecasts. Consequently, we ended up working with the following DLM:

*Observation equation*

$$z_t = s_t^1 + s_t^2 + s_t^4 + s_t^6 + v_t^c, \quad v_t^c \sim N(0, v^c)$$

where $s_t^1$ designates the level of the series; $s_t^2$ and $s_t^3$ refer to first harmonic of the seasonal term; $s_t^4$ and $s_t^5$ refer to the second harmonic and $s_t^6$ refers to the autoregressive term; and $v_t^c$ designates a Gaussian error term of constant, but unknown, variance $v^c$.

*System equation*

$$s_t^1 = s_{t-1}^1 + w_t^{1c}$$
$$s_t^2 = \cos(\pi/6)s_{t-1}^2 + \sin(\pi/6)s_{t-1}^3 + w_t^{2c}$$
$$s_t^3 = -\sin(\pi/6)s_{t-1}^2 + \cos(\pi/6)s_{t-1}^3 + w_t^{3c}$$
$$s_t^4 = \cos(\pi/3)s_{t-1}^4 + \sin(\pi/3)s_{t-1}^5 + w_t^{4c}$$
$$s_t^5 = -\sin(\pi/3)s_{t-1}^4 + \cos(\pi/3)s_{t-1}^5 + w_t^{5c}$$
$$s_t^6 = 0.4s_{t-1}^6 + w_t^{6c}$$

with $\mathbf{w}_t^c = (w_t^{1c}, w_t^{2c}, \ldots, w_t^{6c})$ being an error term such that

$$\mathbf{w}_t^c \sim N\left(0, \begin{pmatrix} v^c W_t^{*c} & 0 \\ 0 & \sigma^{c2} \end{pmatrix}\right)$$

where $\sigma^{c2}$ is the autoregressive variance, and $W_t^{*c}$ is the variance matrix (up to term $v^c$ of the first five terms. This matrix was defined using discounting, with a level discount factor $\delta_1^c$ and a seasonal part discount factor $\delta_2^c$.

*Prior information*

$$\mathbf{s}_0 \mid \phi^c \sim N(\mathbf{m}_0^c, v^c C^{*c})$$

$$\phi^c \sim \mathrm{Gamma}(n_0^c/2, d_0^c/2)$$

with $\mathbf{s}_0 = (s_0^1, s_0^2, \ldots, s_0^6)$ and $\phi^c = 1/v^c$.

The assessment of the gamma parameters was more complicated, and we based it on a guess for the expected variance and graphical representations of the corresponding densities. As a result of the assessments, $\mathbf{m}_0^c$ was specified as $(0.850, -0.779, -0.103, 0.056, 0.144, 0)$,

$$C^{*c} = \begin{pmatrix} 0.09 & 0 & 0 & 0 & 0 & 0 \\ 0 & 0.017 & 0.005 & 0 & 0 & 0 \\ 0 & 0.005 & 0.015 & 0 & 0 & 0 \\ 0 & 0 & 0 & 0.02 & -0.002 & 0 \\ 0 & 0 & 0 & -0.002 & 0.022 & 0 \\ 0 & 0 & 0 & 0 & 0 & 0.1 \end{pmatrix}$$

$n_0^c = 8$ and $d_0^c = 1.2$.

Extensive sensitivity analyses were conducted with respect to $\delta_1^c$ and $\delta_2^c$; the gamma parameters $n_0^c$ and $d_0^c$; the autoregressive coefficient and variance. We studied the effect of changes in the initial estimates of these parameters on the behaviour of predictive variances, mean absolute error of one-step-ahead forecast errors and their autocorrelation functions. The model seemed fairly robust and we selected the following discount parameters: $\delta_1^c = 0.95$ and $\delta_2^c = 0.95$.

### 5.3.3.3   Inflows to Cahora Bassa

The model we propose now for Cahora Bassa depends on the release from Kariba and the incremental inflow $(inc_t)$ to Cahora Bassa from the tributaries and basin between the Kariba gorge and the inlet to Cahora Bassa. Taking into account the fact that water travelling time is less than a month, we use the following relation:

$$i_t^c = u_t^k + inc_t$$

Given that data about releases are not available (and they will actually depend on the releases of our approach), we shall estimate incremental inflows as follows. First, if there was no reservoir, we would have

$$i_t^c = i_t^k + inc_t$$

The second relation describes a dynamic regression between inflows to Cahora and Kariba, and reflects the physical relation that inflow is related to basin size:

$$i_t^c = \beta_t i_t^k$$

Simple computations suggest modelling incremental inflows by

$$inc_t = (\beta_t - 1)i_t^k$$

and inflows to Cahora by

$$i_t^c = u_t^k + (\beta_t - 1)i_t^k$$

Observe that we no longer have a dynamic linear model, since both the regression weights and regressors are subject to uncertainty. Besides, the distribution of $u_t^k$ is not standard. Forecasting $i_t^c$ is easily done by simulation, since we know or can easily sample the distributions involved. That of $i_t^k$ is obtained from the model in Section 5.3.3.1; $u_t^k$ is obtained from Section 5.3.3.1 and using Table 5.3; and the distribution on $\beta_t$ comes from the inversion of the transformation in Section 5.3.3.2 and the model there. After a sufficient number of stages, it will follow a lognormal distribution.

## 5.3.4   Utility functions

Following the description in Section 5.1.2, the utility function used in our approach takes into account the consequences of interest, described in Section 5.3.2, and deviations to a reference trajectory. We describe first how we compute reference trajectories and then the full preference model.

### 5.3.4.1   *Computation of the reference trajectory*

Here we describe the computation of the reference trajectory for Cahora Bassa. A similar procedure was applied in RS to obtain one for Kariba. Essentially, we use deterministic versions of our problems, where inflows are considered known and fixed at their predictive expected values. The dynamics are the same as those of Section 5.3.2. We set the planning period equal to 1 year.

For the deterministic problem, we need an initial volume. We adopt

$$s_0^c = 0.7M^c = 42\ 000 \tag{5.7}$$

The objective function includes a term relating to deviation from the initial state, a term relating to the main objective in the stochastic problem, which is energy production, and a term which deals with separation from the states allowing for fishing (storage between 41 500 and 51 600), defined by the function

$$g(s^c) = \begin{cases} 0 & \text{if } 41\,500 \leqslant s^c \leqslant 51\,600 \\[2ex] \dfrac{1}{2\,141\,400\,000}(s^{c2} - 93\,100 s^c) + 1 & \text{otherwise} \end{cases}$$

Then, the objective function is

$$f(u^c) = \rho_0(s_{12}^c - s_0^c)^2 - \rho_1 \sum_{t=1}^{12} \eta u_{1t}^c(l_t - r_t) + \rho_2 \sum_{t=1}^{12} g(s_t^c)$$

with $\rho_0$, $\rho_1$, $\rho_2$ being positive numbers chosen to give similar weight to the three components and improve numerical stability.

The optimization problem is min $f(u^c)$ s.t. (5.1)–(5.7). We solved it by dynamic programming, with several reformulations to simplify the solution. Each subproblem at each stage was solved with a modified version of OPQSQP (NOC, 1990). A multistart strategy was adopted to avoid bad local optima. For every month $t$, a smooth function was adjusted as an approximation to the optimal value function.

The reference trajectory was obtained as the optimal solution of the above optimization problem and is described in Table 5.7. An additional con-

**Table 5.7**  Reference trajectory for Cahora Bassa

| Month | Initial | Rel. Tur.* | Spillage | Energy | Final |
|---|---|---|---|---|---|
| October | 42 000 | 4848 | 0 | 1052 | 42 256 |
| November | 42 256 | 4848 | 0 | 1053 | 41 420 |
| December | 41 420 | 4848 | 0 | 1049 | 43 293 |
| January | 43 293 | 4848 | 0 | 1047 | 48 258 |
| February | 48 258 | 4848 | 0 | 1063 | 55 948 |
| March | 55 948 | 4848 | 18 000 | 956 | 45 969 |
| April | 45 969 | 4848 | 18 000 | 922 | 32 458 |
| May | 32 458 | 4848 | 0 | 1013 | 35 660 |
| June | 35 666 | 4848 | 0 | 1026 | 38 037 |
| July | 38 037 | 4848 | 0 | 1036 | 40 065 |
| August | 40 065 | 4848 | 0 | 1044 | 41 457 |
| September | 41 457 | 4848 | 0 | 1048 | 42 179 |

*Rel. Tur. = release through turbines.

straint, $u_{2t}^c \leqslant 18\,000$, was introduced to achieve a more homogeneous performance. It corresponds to an excellent reservoir performance, with high energy output and fairly homogeneous releases.

### 5.3.4.2   *The expression of the utility function*

As described in RS, for Kariba we used an additive decomposition for the utility function, because of good first order approximation. Therefore, we include a term relating to deficit and a term relating to spills. The last term is a penalty to deviations from the reference state. The general form is

$$F_K(u_1^k, u_2^k) = F_K(k, u_2^k, s^k) = \lambda f_1(k) + (1 - \lambda)f_2(u_2^k) + \rho_K(s^k - s^{k*})^2$$

where $k$ represents the existence (1) or not (0) of deficit; $s^k$ is the final state of the reservoir; $s^{k*}$ is the reference state; $u_2^k$ is the volume of water spilled; $\lambda$ and $\rho_K$ are weights; and $f_1$ and $f_2$ are component utility functions. For $f_1$, given that $k$ may only attain two values, 0 (no deficit) and 1 (deficit), and that 0 is better, we choose

$$f_1(k) = 1 - k$$

For $f_2$, given that current management is risk averse to large releases and small releases are preferred to large ones, we chose a (risk averse) non-decreasing exponential utility function:

$$f_2(u_2^k) = -0.071\ 71 + 1.083\ 65\ \exp(-0.000\ 1415 u_2^k)$$

We used standard assessment methods (see, e.g., French, 1986) for eliciting the parameters, with $\lambda = 0.75$. We chose $\rho_K = -10^{-10}$ to appropriately scale penalties with other component utilities.

For Cahora Bassa, we modelled

$$F_C(u_1^c, u_2^c) = F_C(E^c, s^c) = f_3(E^c) + \rho_C(s^c - s^{c*})^2$$

where $E^c$ is the energy produced, and $s^c$ and $s^{c*}$ are the storage and reference storage, respectively, at Cahora Bassa. Using standard utility assessment techniques, we used a convex–concave utility function with 765 GWh/month as the inflection point:

$$f_3(E^c) = \begin{cases} 0.0038\,[\,-1 + \exp(0.0066 E^c)] & \text{if } E^c \leq 765 \\ 1.003\ 473 - 0.403\ 4728 \exp(-0.023\ 775\ 71\,(E^c - 765)) & \text{otherwise} \end{cases}$$

For the same reasons above, we used $\rho_C = -10^{-10}$.

Finally, both utility functions are composed so that

$$F(u_1^c, u_2^c, u_1^k, u_2^k) = \mu F_K(u_1^k, u_2^k) + (1 - \mu)F_C(u_1^c, u_2^c)$$

and the expected utility is

$$\Psi(u_1^c, u_2^c, u_1^k, u_2^k) = \int F(u_1^c, u_2^c, u_1^k, u_2^k)\,dH(i^k, i^c \,|\, D)$$

where $H(i^k, i^c \,|\, D)$ is the predictive distribution for the inflows to the reservoirs, obtained from the forecasting model in Section 5.3.3. As initial value for $\mu$, we used 0.5. This and other parameters were checked via a sensitivity analysis.

### 5.3.5   Expected utility maximization

At every time step, the expected utility function had to be maximized with respect to control variables and subject to constraints on the controls (releases from the reservoir): the amount of water released has to be non-negative and the amounts of water released for energy production and flow control are limited by the capacity of the turbines and the spillgates. The optimization problem is thus given as follows:

$$\max \Psi(u_1^c, u_2^c, u_1^k, u_2^k)$$
$$\text{s.t.} \quad 0 \leq u_1^k \leq m^k$$
$$0 \leq u_2^k$$
$$0 \leq u_1^c \leq m^c$$

with the constraints imposed on $u_2^c$ due to the structure of the Cahora Bassa dam, as in eq. (5.5). $m^k$ and $m^c$ designate the maximum releases through Kariba and Cahora Bassa turbines, respectively.

Note, however, that we do not have an explicit expression for the expected utility. The complications from the calculation are reflected in the influence diagram in Fig. 5.4 describing the decision making problem at

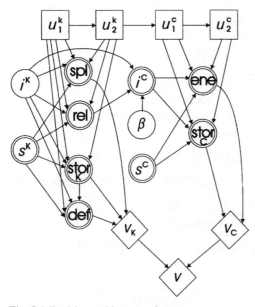

**Fig. 5.4** Decision problem at each stage

each stage. However, for each $(u_1^k, u_2^k, u_1^c, u_2^c)$, the expected utility may be easily computed by simulation. For $(u_1^k, u_2^k, u_1^c, u_2^c)$:

1. Generate $(i_1^k, \ldots, i_N^k)$, from the predictive distribution in Section 5.3.3.1.

2. For each $i_j^k$, and given $(u_1^k, u_2^k, u_1^c, u_2^c)$:

   (i) compute spill, total release, final storage, deficit, $F_K(u_1^k, u_2^k)_j$ (with appropriate tables, like Table 5.3);

(ii)  from the predictive distribution in Sections 5.3.3.1, 5.3.3.2 and 5.3.3.3 and release from Kariba, generate $i_j^c$;

(iii) compute energy, final storage, $F_C(u_1^c, u_2^c)_j$ (with appropriate tables, like Table 5.2).

3.  Approximate $\Psi(u_1^c, u_2^c, u_1^k, u_2^k)$ by

$$\hat{\Psi}_N(u_1^c, u_2^c, u_1^k, u_2^k) = \frac{1}{N} \sum_{j=1}^{N} [\mu F_K(u_1^k, u_2^k)_j + (1 - \mu)F_C(u_1^c, u_2^c)_j]$$

Consequently, at each stage, we solve the problem

$$\max \hat{\Psi}_N(u_1^c, u_2^c, u_1^k, u_2^k)$$

with the constraints above.

We solved those problems, with the Nelder Mead algorithm, as implemented in IMSL (1989), the main reason being the need for a robust method, given the changes in objective function at different stages, and the need for a method requiring only function evaluations. To avoid convergence to bad local optima, we used a guided multistart strategy.

### 5.3.6   Performance

The proposed approach was tested with monthly inflow data from October 1930 to September 1965. Simulations included a warming-up period for the forecasting algorithm, followed by a period in which both forecasting and optimization took place. The first issue we analysed was the impact of key parameters in the performance of the reservoir, mainly $\mu$, $\rho_K$ and $\rho_C$. Sensitivity analysis of other parameters has been described in RS.

First, since spills from Kariba were fairly irregular, $\rho_K$ was modified to $-10^{-9}$, and, for homogeneity, a similar change was made for $\rho_C$. In spite of improvements, we decided to include upper bounds for announced spills from both reservoirs. Effectively, we introduced maximum announced spills of 4000 and 8000 mln m$^3$ for Kariba and Cahora Bassa, respectively, reflecting the desire to control Kariba more tightly, and the relatively less dangerous downstream effects of Cahora spills. Moreover, since at high inflow levels reservoirs tended to be too full, a rule was introduced which increased $\rho_K$ and $\rho_C$ to $-5 \times 10^{-9}$ when reservoirs were at 90 per cent of their maximum storage. $\mu$ was initially fixed at 0.5, and varied in the range $[0, 1]$. The behaviour was fairly stable in the range $[0.4, 0.6]$, whereas for more extreme values, a more irregular behaviour was appreciated. As a consequence, our final simulations were made with our initial value of $\mu = 0.5$. Figures 5.5 and 5.6 provide the behaviour of both reservoirs with our policies, showing releases through turbines, spilled water, energy produced and storage.

As far as Kariba is concerned, note that the energy output is well above the target set of 750 GWh/month, which is much higher than the current target of 600 GWh/month, hence suggesting the potential for a much more efficient operation. Regarding releases, they are always below the set-up

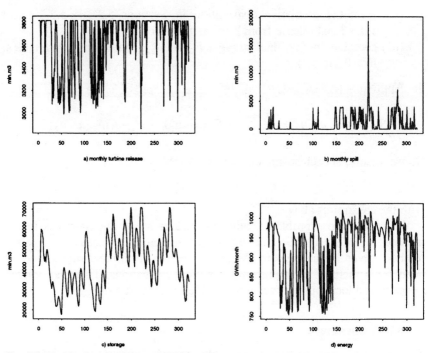

**Fig. 5.5** Simulated performance of Kariba: releases, storage and energy output

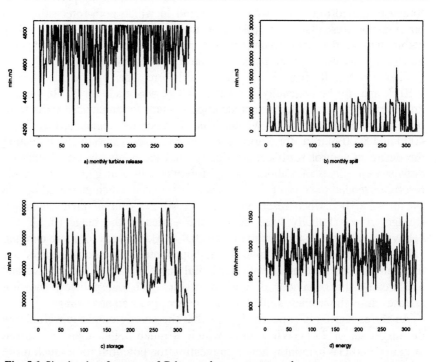

**Fig. 5.6** Simulated performance of Cahora: releases, storage and energy output

level of 4000 mln m$^3$, except for two periods: one of 3 months, in which releases were 19 451, 11 331 and 4491, respectively, and another of 2 months, with spills of 5392 and 7108, respectively. Note that in the first case the inflows in those three months were 25 486, 27 690 and 14 281 (enough to fill up the reservoir!), and in the second case, the inflows in those two months were 15 899 and 10 637, with additional very high inflows (12 295, 17 475) the previous two months, therefore corresponding to a fairly acceptable regulation of the inflow. Globally, the spills are fairly homogeneous and much smaller than those under current management. Also, except for those 5 months in which the reservoir was at maximum retention level, the operation was under safe conditions. Releases through turbines tend to be close to the maximum of 3820 mln m$^3$.

A similar pattern emerges for Cahora Bassa. Releases through turbines tend to be close to the maximum of 4848.76. Spills are smaller than 8000, and are fairly homogeneous, except for 6 months, with spills of 9093 (with inflows the previous 4 months equalling 56 000), 29 302 and 13 539 (with inflows the previous 4 months of 81 000), and 17 506, 15 116 and 9794 (with inflows the previous six months of 106 000). Again, we operate at fairly safe levels regulating inflows satisfactorily. The energy output ranges between 895 and 1066 GWh/month. The operation of both reservoirs is fairly well balanced, both in economic and safety terms.

We tested several scenarios. First, we studied performance of the reservoirs starting from very low storages. Initially, the energy production was not very high, in an attempt to fill up the reservoir progressively, until high enough storages were achieved, after which the performance was good enough. Similarly, we tested the operation of the reservoir for initial storages close to their maxima: again, the reservoir gradually reached safer (lower) levels. In this case, adaptation was much faster than that described in RS, given the rule for high storages introduced.

Finally, we simulated the reservoir with very low inflows. Since the real inflows available were not low, we generated inflows artificially by dividing the real ones by 4. Under these extreme conditions the performance was still acceptable, in the sense of there being releases only for energy production and only so as to satisfy energy targets. Obviously, those settings are too high for very low inflows, suggesting the need to change targets under these conditions or to reformulate objectives and preferences.

The typical shape of the release rules obtained as the result of the computations is as follows: for low values of storage, the amount of water released for energy production decreases as the storage increases, since smaller amounts of water are required to achieve a given energy production target and we want to remain close to the reference state; at high storage levels, the operating policy tends to release as much water as possible to generate electricity, minimizing spills. For very high storage levels, the mechanism controlling deviation of the current state from the reference values 'turns on' and excessive water is spilled in order to secure that the reference trajectory is followed (Fig. 5.7).

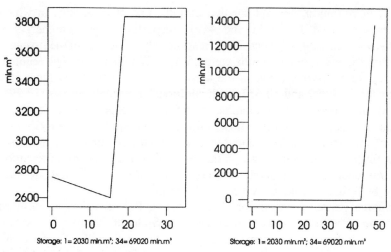

**Fig. 5.7** Typical shape of policies

## 5.4   Discussion

Various reviews (e.g., Yakowitz, 1982; Yeh, 1985) of reservoir operations describe multireservoir operation problems under uncertainty as extremely complex. In our case, there were additional complications due to the presence of multiple objectives. We have shown that our approach may successfully cope with this kind of problem.

Key issues are the definition of flexible policies; the use of Bayesian forecasting models; a careful modelling of preferences, which include a term reflecting deviation from a reference trajectory; a heuristic providing policies of approximate maximum expected utility; and thorough checking of our policies through sensitivity analyses, to provide additional modelling insights.

We would like to stress that a basic principle underlying our approach is that of *management by exception* (West and Harrison, 1997). The central idea is that we use a set of models to process information, and make predictions and decisions, unless exceptional circumstances arise, in which subjective intervention is required. This may be reflected in various ways. For example, the forecasting model could require interventions if sudden rainfall suggests an increase in the inflow level. It is also reflected in the use of predictive expected inflows in the definition of the reference trajectory, with obvious differences from wet to dry years. Finally, it is also reflected in the rule introduced to handle cases in which reservoir storages are very high. Obviously, other types of interventions would be permitted, making our approach truly interactive.

This case study and our previous one (RS) suggest the enormous potential of our approach in handling reservoir operation problems and, more generally, multistage decision problems under uncertainty. However, its application is far from simple. We are currently building Bayres (Ríos Insua *et al.*, 1996), a decision support system for reservoir operations that would support all the phases of our approach. On a more theoretical basis, we are evaluating the quality of our heuristic in general multistage decision problems. Other applications of Bayesian methods in water resources problems are described in Berger and Ríos Insua (1996).

# References

Balek, J. 1977: *Hidrology and water resources in tropical Africa*. Development in Water Science, 8. Amsterdam: Elsevier.

Berger, J. O. and Ríos Insua, D. 1996: Recent developments in Bayesian Inference with applications in Hydrology. *Technical Report*, Dpt of AI, Madrid Technical University.

French, S. 1986: *Decision theory*. Chichester: Ellis Horwood.

Gandolfi, C. and Salewicz, K. 1990: Multiobjective operation of Zambezi river reservoirs. *IIASA WP-90-31*. Laxenburg, Austria: IIASA.

Gandolfi, C. and Salewicz, K. 1991: Water resources management in the Zambezi Valley: analysis of the Lake Kariba operation. In Van De Ven, F. H. M., Gutknecht, D. and Salewicz, K. A. (eds) *Hydrology for the management of large river basins*. IAHS Publication No. 201, 13–25. Wallingford, UK: Institute of Hydrology.

IMSL 1989: *IMSL math/library, user's manual*. Houston, TX: IMSL, Problem-Solving Software Systems.

Keeney, R. L. 1992: On the foundations of prescriptive decision analysis. In Edwards, W. (ed.) *Utility theories: measurement and applications*. Boston: Kluwer.

Klemes, V. 1981: Applied stochastic theory of storage in evolution. *Advances in Hydrosciences* **12**, 79–141.

Lamond, B., Monroe, S. and Sobel, M. 1995: A reservoir hydroelectric system: exactly and approximately optimal policies. *European Journal of Operational Research* **81**, 535–42.

Loucks, D. P. and Sigvaldasson, O. T. 1982: Multiple reservoir operation in North America. In Kaczmarek, Z. and Kindler, J. (eds) *The operation of multiple reservoir systems*. IIASA CP-82-S02. Laxenburg, Austria: IIASA.

NOC 1990: OPQSQP. Hatfield Polytechnic: Numerical Optimisation Centre.

Pinay, G. 1988: Hydrobiological assessment of the Zambezi. *IIASA WP-88-089*. Laxenburg, Austria: IIASA.

Ríos Insua, D., Bielza, C., Martín, J. and Salewicz, K. 1997: BayRes: a system for stochastic multiobjective reservoir operations. To appear in Proceedings of the Second International Conference in Multi-objective Programming and Global Programming. Springer Verlag.

Ríos Insua, D. and Salewicz, K. 1995: The operation of Kariba Lake. *Journal of Multicriteria Decision Analysis* **4**, 203–22.

Salewicz, K. and Loucks, D. 1988: Decision support systems for managing large international rivers. *Interim report to the Ford Foundation*. Laxenburg, Austria: IIASA.

Salewicz, K., Loucks, D. and McDonald, A. 1989: Decision support systems for managing large international rivers. *Interim Report to the Ford Foundation*. Laxenburg, Austria: IIASA.

UNEP 1987: *Agreement on the action plan for the environmentally sound management of common Zambezi river system: final act*, Harare 26–28 May 1987. Nairobi: UNEP.

Venema, H. D. and Schiller, E. J. 1995: Water resources planning for the Senegal river basin. *Water International* **20**, 61–71.

West, M. 1995: Bayesian forecasting, *ISDS Discussion Paper*. Durham, NC: Duke University.

West, M. and Harrison, J. 1997: *Bayesian forecasting and dynamic linear models*. New York: Springer.

Yakowitz, S. 1982: Dynamic programming applications in water resources. *Water Resources Research* **18**, 673–96.

Yeh, W. 1985: Reservoir management and operations models: a state-of-the-art review. *Water Resources Research* **21**, 1797–818.

# 6 Event conditional attribute modelling in decision making when there is a threat of a nuclear accident[1]

S French, M T Harrison, D C Ranyard

## 6.1 Introduction

Major nuclear accidents such as those at Three Mile Island and Chernobyl have directed attention to the need for tools to support coherent decision making on protective actions. Within the European Union one response has been the development of RODOS,[2] a decision support system (DSS) for nuclear emergency management. This is being designed and built by a consortium of European institutes in association with several eastern European and former Soviet Union collaborators. Here we focus on supporting decision making in the pre-release phase when there is a threat, but as yet no actual release of radioactivity. Decision makers have to balance many conflicting objectives, knowing that the consequences of their decisions will be judged with hindsight very differently if an accident happens than if it does not.

Decision making on measures to protect the public in the event of a

[1] This work is being carried out as part of the RODOS programme of research funded by the EU (F13P-CT92-0036, F13P-CT92-013b, Sub94-F15-028 and Sub94-F15-045). The views expressed in this paper are those of the authors and do not necessarily reflect those of the RODOS project. We are grateful for many discussions with M. Ahlbrecht, J. Ehrhardt, G.N. Kelly, M. Morrey, J.Q. Smith and many other collaborators on the RODOS project. Discussions with Larry Phillips and T.J. Stewart were also very useful.

The example described in Section 6.5 was built using DPL (ADA Decision Systems, Inc., 2710 Sand Hills Road, Menlo Park, CA 94025, USA).
[2] Real-time Online DecisiOn Support system for nuclear emergencies. A more accurate title for the project, 'Programmed Analysis of Nuclear Incidents and Countermeasures', was suggested by Jim Smith and myself, but was not adopted!

release of radioactivity is complex and involves many criteria. Aside from direct health effects, psychological stress and public acceptability must be taken into account. In the later phases of an accident when a release is known to have occurred, it has proved possible to build attribute hierarchies upon which the decision makers have been able to articulate their judgements of societal values (French, 1996a; Lochard *et al.*, 1992). However in the case of precautionary measures it has proved exceptionally difficult to structure the model. In this paper we propose using event conditional attribute modelling and explore its use in a simplified hypothetical analysis.

In the next section we introduce the context of the problem in greater detail. Subsequently we discuss the issues which seem to mitigate against the standard approach to structuring attribute hierarchies as discussed in, *inter alia*, Keeney (1992). In Section 6.4 we suggest a general theory for event conditional attribute modelling and in Section 6.5 we explore its use in a hypothetical example. Section 6.6 contains further discussion and some concluding remarks.

Further background on nuclear emergency management in general and RODOS in particular may be found in: Ehrhardt *et al.* (1993), French (1996a), French *et al.* (1995a,b), Kelly and Ehrhardt (1995), Walmod-Larsen (1995) and *Radiation Protection Dosimetry* (1993, **50**, issues 2–4).

## 6.2   The context of emergency management

The issues faced in nuclear emergency management are difficult and the political processes involved are complicated. In the course of an incident, responsibility passes between several different groups of decision makers of differing technical and political sophistication. Moreover, the process varies from country to country and from culture to culture. Thus the thumbnail sketch given here should not be taken as definitive. But we hope that it is sufficient to set the scene.

During the building and operation of nuclear plants, many preparations are made to deal with potential emergencies. Data bases of demographic, geographic, agricultural and economic data are established. Evacuation routes and procedures are planned. Exercises are held regularly to practise different accident scenarios. But no accident ever goes 'as planned'. Emergency management is not simply the implementation of predetermined rules: in the event of an accident there are many decisions to be made. Emergency managers are provided with national and international guidance in the form of *intervention levels* on when to use each type of counter-measure, e.g. sheltering, evacuation or food bans. Generally, these are lower and upper levels on the predicted exposure. Below the lower level the advice is that intervention is unnecessary, whereas above the upper level it is required. Between the lower and upper levels, the action to be taken is left to the discretion of the emergency management team in the light of the particular circumstances.

When there is a threat of an accident, plant managers would take appropriate engineering actions to avoid or reduce the risk of a release. The first decisions on protecting the public would be whether to take precautionary measures[3] such as warning the public; distribution of stable iodine tablets; and beginning the evacuation of some areas. If a release actually occurred, decisions would be needed on protective actions such as (issue and) uptake of stable iodine; advice on sheltering; and evacuation. In the following days further decisions will be needed on such measures as food bans; decontamination of livestock, agricultural produce and properties; and restrictions on business, leisure activities and access to the region. After several days or maybe weeks, decisions will be needed on longer term measures, e.g. permanent relocation and permanent changes to agricultural practice and local industry. We focus entirely on decision making in the early phase when there is a threat, but as yet no actual release.

Before exploring the decision analytic issues, it will be helpful to say a little about how the direct radiation related health effects are assessed. Very briefly: current medical understanding of the human health effects resulting from exposure to radioactivity is as follows. If the exposure is sufficiently high, deterministic health effects may result. A few tens of workers received such high doses during the Chernobyl accident. Our concern will not be with such exposures: we shall assume that the potential accident does not threaten deterministic effects to the public. Rather we consider the health effects arising from the 'low level' increase over background radiation which would be suffered by the majority of the affected population in the event of an accident. Such effects are stochastic. Exposed individuals bear a higher cancer risk and there are risks of genetic effects in subsequent generations. These risks vary according to a number of factors related to the exposure. Firstly, the radiation may be alpha, beta or gamma of different energies and spectra. Secondly, the risk varies according to the organs exposed. Internal exposure arising from inhalation or ingestion is usually more serious than external exposure; and different organs have different susceptibilities. For an individual in a particular set of circumstances for a particular time period, it is possible to calculate from predictions of the spread of contamination a quantity known as the *individual effective dose equivalent*, often shortened to *individual dose*. For low level exposures, current medical belief supported by empirical studies is that the health risks to the individual are linearly related to the individual dose. This is known as the *linearity hypothesis*.

Current practice in radiation protection decision making focuses on the *collective effective dose* (or *collective dose*). This is the sum of the individual doses over a population. Under the linearity hypothesis the expected number of health effects arising from the exposure to a population is proportional to the collective dose. Thus minimizing collective dose is an important criterion, arguably the most important criterion, in such decision making. It should be noted that typically decision makers examine the

---

[3] We use the terms *protective actions* and *(counter)measures* synonymously. *Precautionary measures* refer to those actions taken (or decided upon) before any release has occurred.

collective dose for different subpopulations, e.g. infants under 1 year, children, women of childbearing age, and other adults. Also, they usually distinguish between public and worker populations and may apply some constraints to the maximum acceptable individual dose.

Individual doses are measured in units of Sieverts (Sv): roughly 1Sv increases the probability of a fatal cancer for an individual by 5 per cent, i.e. a lifetime cancer risk of 20 per cent would increase to 25 per cent. Collective doses are measured in ManSieverts: an exposure of 1mSv to each member of a population of 100 000 gives a collective dose of 100 ManSievert and would increase the expected number of cancers in the lifetime of that population by 5.

## 6.3    Attribute modelling for decisions on precautionary measures

RODOS is designed to be a comprehensive DSS for off-site emergency management. It is unique among DSS'S for nuclear emergencies in seeking to provide decision support at all levels ranging from the largely descriptive to providing a detailed evaluation of the benefits and disadvantages of various countermeasure strategies and ranking them according to the societal preferences as perceived by the decision makers (*see* Table 6.1). The design of RODOS makes assumptions about the needs of decision makers during level 3 support. However, there is little practical experience of these needs, particularly in relation to decision-making on precautionary and early countermeasures. Moreover, in order to offer real-time support in the short time-scales necessary in the early phases of an accident there is a need to

**Table 6.1** Decision support can be provided at various levels, here categorized into four levels. The functions provided at any level include those at lower levels. RODOS will provide support at all levels, including level 3 for all potentially useful countermeasures at all times following an accident. (After Kelly, 1994.)

| Level | Function |
| --- | --- |
| Level 0 | Acquisition and checking of radiological data and their presentation, directly or with minimal analysis, to decision makers, along with geographic and demographic information available in a geographic information system |
| Level 1 | Analysis and prediction of the current and future radiological situation (i.e. the distribution over space and time in the absence of countermeasures) based upon monitoring date and meteorological data and models |
| Level 2 | Simulation of potential countermeasures (e.g. sheltering, evacuation, issue of iodine tablets, food bans, and relocation), in particular determination of their feasibility and quantification of their benefits and disadvantages |
| Level 3 | Evaluation and ranking of alternative countermeasure strategies in the face of uncertainty by balancing their respective benefits and disadvantages (e.g. costs, averted dose, stress reduction, social and political acceptability), taking account of societal preferences as perceived by decision makers |

prime the system with default value judgements. These defaults need to be elicited. Experience within decision analysis suggests that such elicitation cannot occur in the abstract. To be useful, judgements must always be elicited within a realistic context (Keeney, 1992). As a first step, a series of exercises using the prototype RODOS system are being run. These involve senior decision makers in desktop simulations in which they face the possibility, but not the certainty, of an imminent accidental release. Of particular interest are:

- the attributes considered to be important in evaluating countermeasures;

- whether these attributes are evaluated linearly or whether they need to be transformed in a non-linear sense to encode, e.g., risk attitude;

- the weights the decision makers ascribe to these attributes;

- the attitude of the decision makers to information when it is provided with a statement of the uncertainty involved.

For discussions of the use of multi-attribute modelling techniques, see French (1986, 1996a), Keeney and Raiffa (1976) and Keeney (1992).

It is too early to report firm results from these exercises. Some are still to be run, and those that have been run are still being analysed. Moreover, there are confidentiality matters to be resolved, particularly since we are comparing attitudes in different member states of the European Union. However, several points are clear.

We expressed the uncertainty relating to the threat of an accident probabilistically. This went beyond the decision makers' experience and the introduction *per se* of uncertainty was discomforting for the decision makers. Most exercises to date have rehearsed emergency management issues under the assumption that a release is certain to happen or has happened. To a certain extent, this meant that the decision makers had been trained equate 'probability of release' with 'certainty of release'.

There are national and international guidelines which suggest that countermeasure strategies should be evaluated in terms of the dose that they avert.[4] Precautionary measures may avert no dose because there may be no release. Moreover, the expected averted dose may not exceed any intervention level because the probability of a release is too low. Thus it is unclear how to evaluate such measures in a manner consistent with averted dose methodologies. Some groups tackled this problem by assuming the worst was certain to happen and by taking the protective actions required by their national intervention levels. Others argued that there was no release as yet and therefore no action should be taken to protect the public immediately, although many preparations should be made in case it were necessary.

Equity issues were perceived as very important. The public have a right to expect fair and equal protection in the event of an accidental release of

---

[4] There are clear parallels between using averted dose as a decision criterion in this context and Savage's regret measure (Savage, 1954; French, 1986).

radiation. However, 'equity' comprises many concepts:

- *ex ante equity* – the equity of the process and risks which eventually lead to the health effects;

- *ex post equity* – the equity associated with the health effects that actually occur;

- *dispersive equity* – the equity of the distribution of risk over groups within the population. One can also discuss *ex ante* and *ex post* dispersive equity.

Taken alone, each of these is difficult to articulate within a decision model. In this case the problems are compounded because, when the focus is on precautionary measures, there is a very real possibility of an accident and, hence, a move from *ex ante* to *ex post* equity considerations. The decision makers are aware that *ex ante* equity issues will be discussed with hindsight in the event that there is no release, whereas *ex post* equity issues will replace them in the event that there is one. See French *et al.* (1995) for a survey and more detailed discussion of equity issues in this context.

It has proved far more difficult to develop attribute hierarchies for these situations than we imagined. In fact, in the four exercises held to date, we have been unable to build complete hierarchies. Some attributes have been identified, but in no case has a completely adequate or acceptable hierarchy been built.

The reference to *hindsight* immediately above is significant, because it points to a key difficulty facing the decision makers in their thinking about precautionary measures. They have to balance many conflicting objectives, knowing that the consequences of their decisions will be judged with hindsight very differently if an accident happens than if it does not. In short, one needs to consider aspects of *anticipated regret* (Janis and Mann, 1977). Anticipated regret has been much discussed in the psychological literature (see, e.g. Richard *et al.*, 1996). It can lead to decision making behaviours that are less than rational (Russo and Schoemaker, 1989). There have been many attempts to provide normative models of 'rational' regret; see, e.g., Bell (1982), Loomes and Sugden (1982), and Savage (1954).[5] These model the regret that decision makers feel if they learn that they could have done better had they taken another course of action. In some cases this is achieved by a comparison of the numerical evaluation of the consequence of the action chosen compared with that arising from the best possible action chosen with hindsight. Others introduce additional attributes which seek to model regret. French (1996b) argues in favour of the latter approach. Here

---

[5] Earlier we noted parallels between Savage's regret model and the use of averted dose. The affect of anticipatory regret and its introduction in the decision model as discussed here is slightly different. Averted dose tries to capture the idea that 'Given the situation we could not do anything more cost-effective'. Anticipatory regret seeks to capture the anticipation of the emotion 'If only we had done that, things would have been so much better'. There is a much stronger emotional emphasis.

we extend it – in a sense – by using different attribute trees to articulate the decision makers' value judgements in the case that a release occurs and in the case that it does not. This approach avoids the introduction of 'regret' attributes since it may be inappropriate to articulate these explicitly in emergency management conducted by government authorities, but does allow the modelling of the decision makers' evaluation of the radically different world after a release has occurred from that after a release has been avoided.

## 6.4   Event conditional attribute modelling

The imperative to support *consistency* in decision making is central to the motivation of Bayesian decision analytic methods. One of its manifestations is apparent throughout the literature on multi-attribute value and utility modelling: see, e.g., French (1986), Keeney and Raiffa (1976) and Keeney (1992). Decision analysts seek to use the *same* attributes to describe all the consequences within a decision model. By doing so they ensure that all consequences are compared along the same dimensions, i.e. there is a consistency in the comparison. An exception may be found in French and Vassiloglou (1983), which considered educational assessment issues arising when students are required to be graded on a common scale having taken different options in their studies. Also there have been multi-attribute resource allocation analyses in which some alternatives have been described against different sets of attributes (Phillips, 1995).

None the less, the overwhelming majority of applications of multi-attribute modelling and the entire thrust of the theoretical literature emphasize the need to use a common set of attributes. However, that need does not arise within the axiomatic foundations of Bayesian decision analysis. It arises because it reduces the number of judgements required of the decision makers in exploring the appropriate form of multi-attribute value or utility model. In our application, there is evidence that the decision makers find the cognitive demands that the drive for a common set of attributes places on them too great. Thus the balance of cognitive effort may be in favour of a process which demands more, but potentially simpler, judgements to assess a value or utility model.

The axiomatic base of subjective expected utility (SEU) begins from a description of a decision problem based upon a set of states, $\Theta$, a set of alternatives, $A$, and a set of consequences, $C$ (French, 1986, 1997). The development of the model which ranks the alternatives according to expected utilities requires, *inter alia*, that the decision makers can – or, from a prescriptive viewpoint, should be able to – provide a weak order representing their preferences between the consequences. To do this the decision makers must understand – i.e. have an image in their mind of – each of the consequences. Thus $C$ must be a set of *requisite* descriptions of consequences, where by 'requisite' we mean that the descriptions are sufficiently detailed for the decision makers to have confidence in their ranking (Phillips, 1984). The development of the utility function, $u: C \rightarrow \Re$, requires no more structure than this in $C$. However, the assessment of $u(.)$ when $C$ has such minimal properties requires that the decision makers express very many preferences and indifferences between hypothetical lotteries. If $\#C = n$, then the minimum number of such judgements is $(n - 2)$. For infinite (or large) $n$ this may be quite time-consuming! Thus one looks to introducing structure into $C$ which may help the decision makers shorten the assessment process, e.g. topological and convexity assumptions may be introduced to allow the risk attitude to be defined and modelled. This can greatly shorten the assessment through the 'smoothing in' of an utility curve, particularly when supported by sensitivity analysis (Keeney and Raiffa, 1976, p. 153–7).

In multi-attribute utility modelling, structure is introduced by requiring $C$ to be a subset of $\Re^q$, in which each of the dimensions is an *attribute* which describes the consequence in terms related to a distinct criterion which matters to the decision makers. Independence conditions such as preferential, utility and additive independence may then be used in appropriate circumstances to simplify the functional form of $u: C \subset \Re^q \rightarrow \Re$. For instance, $u(.)$ may be additive; see, e.g., Keeney and Raiffa (1976) or Keeney (1992) for details. The simplification eases the assessment process because the decision makers can concentrate on each attribute in turn and then construct the overall multi-attribute utility from the marginal utility functions.

Similar simplification is available if one assumes $C = C_1 \cup C_2$, where $C_1 \subset \Re^r$, $C_2 \subset \Re^s$, and $\phi = C_1 \cap C_2$, i.e. that $C$ is partitioned into two subsets of different dimensions. Moreover, suppose that the partition given by $C_1$ and $C_2$ is related to the occurrence or non-occurrence of an event in the state space $\Theta$. In fact, $C$ might be partitioned into more than two subsets, but two will serve our application and the generalization is straightforward. To give some substance to this idea, consider Fig. 6.1. The attribute hierarchies correspond to partitioning the consequence space into a four-dimensional

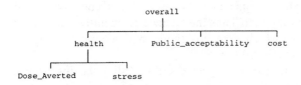

Attribute tree in the event of a release

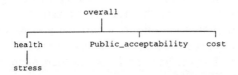

Attribute tree in the event of no release.

**Fig. 6.1** Possible event conditional attribute trees for decisions on precautionary measures

space, $C_1$, in the case that a release occurs, and a three-dimensional space, $C_2$, in the case that one does not. Thus:

$$(\text{Dose\_Averted, stress, public\_acceptability, cost}) \in C_1$$

or

$$(\text{stress, public\_acceptability, cost}) \in C_2$$

depending on whether a release occurs. Note that the attributes 'stress' may have different interpretations (and importance) in the two parts of this decomposition. In the event an accident happens, stress will be very long-lived, with many members of the public fearing for their health for the rest of their and their children's lives. In the event that an accident is averted, stress will be mainly short-lived, although those living close to the plant may feel more vulnerable in the future. It should be emphasized that the hierarchies here are offered as examples only. Other attributes, such as 'potential_for_litigation', may be introduced in practice.

Multi-attribute utility functions $u_1(.)$ and $u_2(.)$ may now be assessed on each of $C_1$ and $C_2$. It will be necessary to bring the two utility functions to a common scale. This will require that the decision makers are asked for a judgement of indifference which relates consequences in $C_1$ to consequences in $C_2$.

To confirm this, let $u: C \to \Re$, $u_1: C_1 \to \Re$, and $u_2: C_2 \to \Re$. Then, since $u(.)$ and $u_i(.)$ represent the same preferences on $C_i$, $u(.) = \alpha_i.u_i(.) + \beta_i$. Without loss of generality we may take $\alpha_1 = 1$, $\beta_1 = 0$. Moreover, on taking expectations to rank alternative strategies, the additive constant $\beta_2$ enters equally on both sides of any comparison, and therefore cancels. Thus to determine $\alpha_2$ one indifference is sufficient.[6] Notwithstanding, it would be sensible to ask for more than one indifference as a consistency check. In this application, these will almost certainly require hypothetical lotteries since the consequences are more serious when there is a release than when there is not: preferences on $C_1$ may not overlap[7] those on $C_2$.

We do not pretend that these judgements will be easy to make, but decision makers do not find the situation without this decomposition at all easy either. As we have noted, they have difficulty in articulating their value judgements in such a way that we can construct a single multi-attribute representation. It might seem tempting to side-step the problem by noting that 'cost' is an attribute in both parts of the decomposition. Surely monetary values will not change depending on the occurrence or not of a release? In fact, it is quite possible that they might. A government is much more likely to argue that 'money is no object' (by which they will mean 'money is little

---

[6] A similar argument shows that if $C$ is partitioned into $r$ subsets corresponding to the consequences which may arise in each of $r$ disjoint events, then $(r-1)$ judgements of indifference will be required. Firstly, $\alpha_1$ can be set arbitrarily to 1; the $\beta_i$ are 'irrelevant constants' which, after taking expectations, cancel out of ranking comparisons; which leave $(r-1)$ $\alpha s$ to be determined.

[7] Viz. $c < c'$, $\forall c \in C_1$, $c' \in C_2$, where $<$ means 'is strictly less preferred to'.

object') if there is a serious accident than if one is averted. Money may become freely available from a contingency fund in the former case and be limited from tight running budgets in the second. Thus the weight on the cost attribute may differ substantially between the two cases.

## 6.5   An example

### 6.5.1   Context of the example

In this section we give an example which we hope will give substance to our discussion. Since we are interested in exploring the modelling of the decision makers' preferences rather than the predictions of the spread of contamination, we used a very simplistic representation of an accident. For instance, we consider just three possible levels of dose rather than a continuous distribution. Fig. 6.2 shows a schematic map of the environs of

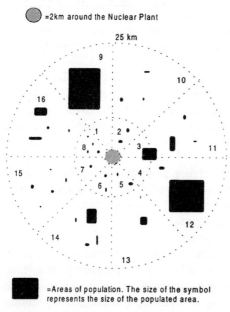

**Fig. 6.2** Hypothetical accident scenario

the hypothetical plant, but note that we use this merely to give context for the hypothetical judgements, rather than numerical detail. Should the methodology proposed here be adopted within RODOS or another DSS, the physical scenario would be much more carefully modelled using full atmospheric and aquatic transport and deposition models and a geographic information system.

We assume that in emergency management planning it has been decided to divide the region 25 km around the plant into sectors as in Fig. 6.2.

Within each sector a *single* package of countermeasures would be applied. In reality the boundaries of the sectors would be modified in the light of geographic and other features. The issue for the decision makers is to decide on appropriate precautionary actions in the face of a threatening nuclear accident. We shall make a further simplification in that we shall assume that precautionary actions may be chosen for one sector independently of decisions on actions elsewhere. This is very unlikely to be the case, and other work within the RODOS project is investigating constraint management techniques to introduce interdependencies between the packages of countermeasures applied in each sector (French, 1996a).

For each sector, there are a number of possible precautionary actions:

- *None*

- *Warn* – the public may be warned so that they are able to prepare for the possibility of evacuation, sheltering and so on. Note that this may lead to some immediate self-evacuation, as may any other precautionary measure.

- *Iodine* – distribution of stable iodine tablets with advice to take them in the event of a release.

- *Shelter* – the public may be told to prepare to shelter.

- *Evacuate* – evacuation should begin as soon as possible, with a mix of self-evacuation and transport provision by the authorities.

In practice, combinations of actions are possible, e.g. the issue of iodine tablets and the advice to prepare for sheltering. For simplicity, however, we consider neither these nor other possible actions. Constraint management techniques in the design of RODOS will identify sensible and feasible combinations for evaluation (French, 1996a).

Fig. 6.3 provides an influence diagram[8] which describes the probabilistic structure of the example. The decision makers' judgemental evaluation of the countermeasures is assumed to be determined by the attribute hierarchies in Fig. 6.1. The four attributes of concern, viz. 'Dose_Averted', 'stress', 'public_acceptability' and 'cost' are shown across the centre of the influence diagram. Note that the values of the attributes of 'Dose_Averted', 'stress' and 'public_acceptability' are influenced by the chance node 'Release?', i.e. by whether an accident happens. This allows the modelling of the use of a different attribute hierarchy in each case. The financial costs of the precautionary countermeasures are assumed to be independent of 'Release?'; note, however, that the weight applied to the cost is not so independent (see below).

The chance node 'Dose' reflects the uncertainties on the size of the source term and wind field. The chance node 'Time Until' moderates the effect of

---

[8] The model is built using DPL software (ADA, 1996). This allows simultaneous decision tree and influence diagram views of the problem. For simplicity, we have suppressed random nodes and dependencies which used to model wind direction and source term.

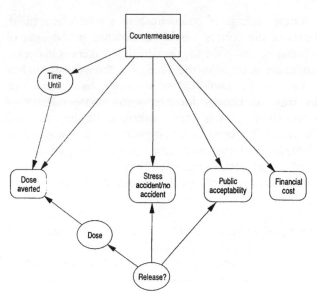

**Fig. 6.3** Influence diagram showing probabilistic dependencies included in the model

countermeasures. The longer the period of warning before a release, the greater the fraction of a population that can be evacuated. Thus the level of the attribute 'Dose_Averted' depends upon the potential 'Dose' and the 'Time Until' any accident occurs. It would also depend on the time of day that an accident occurs. Generally it is easier to find people during the evening or at night than it is during the day and thus implement sheltering or evacuation. The time of day has been factored into the 'Time Until' variable.

Fig. 6.4 shows the decision tree representation of the problem. The decision makers have to decide upon which of the five countermeasures to implement. Then three key uncertainties will be resolved: whether there is an accident, whether it will occur in 6, 10 or 14 h and what the potential dose to the public will be.

In the example we assume that the wind is generally blowing from ESE to SE and that its speed is between 5 and 10 knots. We consider three of the

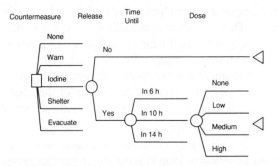

**Fig. 6.4** The decision tree representing the sequence of events in the example

areas in Fig. 6.2: areas 8, 9 and 10. These are representative of the more likely regions to be exposed to the plume. The costs and potential collective dose recognize the much larger population in area 9. Equally the dose calculations recognize that higher doses are likely in area 8 because of its greater proximity to the release. Thus in area 8 the individual dose would be in the range 0.6–3.0 mSv, whereas in areas 9 and 10 the range was 0.3–1.4 mSv. The collective doses in each of the three areas were calculated assuming populations of 400, 20 000 and 500, respectively.

## 6.5.2   Utility judgements

The utilities for the attribute stress are given in Table 6.2. These represent judgements such as that taking no precautionary action when an accident subsequently occurs will cause maximum stress to the public, while taking

**Table 6.2**   Utilities for 'stress'. Note that higher values equal greater preference

| Countermeasure | Accident | No accident |
|---|---|---|
| None | 0.0 | 1.0 |
| Warn | 0.2 | 0.8 |
| Iodine | 0.4 | 0.4 |
| Shelter | 0.5 | 0.8 |
| Evacuate | 0.8 | 0.05 |

no action when an accident is subsequently averted creates the least stress. The values given are for each individual in the affected population. We assume that utility for the stress induced by a countermeasure is the value in the table multiplied by the size of the population.

The 'public_acceptability' of any action is taken to be independent of the population affected since it is interpreted as reflecting the reaction of all the public both in the affected areas and elsewhere in the country. Table 6.3 gives the utilities for public acceptability used in the analysis. Again, higher values correspond with increasing preference.

**Table 6.3**   Utilities for 'public_acceptability'

| Countermeasure | Accident | No accident |
|---|---|---|
| None | 0.0 | 0.8 |
| Warn | 0.25 | 1.0 |
| Iodine | 0.6 | 0.2 |
| Shelter | 0.5 | 0.4 |
| Evacuate | 1.0 | 0.0 |

The cost structure used is given in Table 6.4. It is assumed that the cost of implementing any of the countermeasures is independent of whether there is an

**Table 6.4**   Costs of countermeasures

| Countermeasure | Cost (£1000s) |
|---|---|
| None | 0 |
| Warn | 20 + population*0.001 |
| Iodine | 20 + population*0.010 |
| Shelter | 20 + population*0.005 |
| Evacuate | 20 + population*0.100 |

accidental release. If any precautionary measure is applied, there is a £20 000 fixed cost plus a variable cost depending on the population in the sector. We assume that the marginal utility for the cost is linear. However, in the base case of the analysis we gave cost zero weight in the event that there was a release, reflecting the judgement that money would be 'no object' in such circumstances.

The utility function used in the event of an accident was:

$$u_1(\text{Dose\_Averted, stress, public\_acceptability, cost})$$

$$= x_1 \times \text{Dose\_Averted} + x_2 \times \text{stress} \times \text{population}$$

$$+ x_3 \times \text{public\_acceptability} + x_4 \times \text{cost}$$

while in the event that an accident is averted, the utility function was:

$$u_2(\text{stress, public\_acceptability, cost})$$

$$= a_2(y_1 \times \text{stress} + y_2 \times \text{public\_acceptability} \times \text{population}$$

$$+ y_3 \times \text{cost})$$

The values of the weights in these formulae are given in Table 6.5. The weights $x_1$ and $x_2$ were set according to a judgement that the stress effect for

**Table 6.5**   Values of weights used in the analysis

| $x_1$ | $x_2$ | $x_3$ | $x_4$ | $a_2$ | $y_1$ | $y_2$ | $y_3$ |
|---|---|---|---|---|---|---|---|
| 1.0 | 1.0 | 100.0 | 0.0 | 0.1 | 1.0 | 100.0 | 0.1 |

an individual would be of the same order as 1 mSv of dose. The weight $x_3$ was then set by the judgement that public acceptability issues might begin to dominate in dealing with populations of order 100 or less, but in larger populations the balance between health and dose effects should dominate. We set $y_1 = x_2$ and $y_2 = x_3$ to represent a similar judgements about the relative importance of stress and public acceptability. Although the weights

$y_i$ apply in the event of no release, if one chains the argument through, the weight $y_3 = 0.1$ corresponds to the monetary value of averting a ManSievert being about £100 000.

Complete DPL listings for the example are available from Simon French. From these it is possible to see the complete data sets.

### 6.5.3 Elicitation of $a_2$

For the purposes of the analysis described here, we assume that the utility function is linear, albeit with two forms depending on the occurrence of an accident (see above). This means that the values of the marginal utility functions for stress and public acceptability may be assessed by any of the standard methods discussed, *inter alia*, in Keeney and Raiffa (1976). The weights $x_i$ and $y_i$ can be assessed by the swing-weighting or lotteries techniques. However, some thought needs to be given to how $a_2$ may be assessed.

One method might be to discuss the monetary value of ManSievert in normal circumstances and post-accident. There has been considerable debate in the radiological protection community about the relative monetary values of a ManSievert which should be used in siting decisions and emergency management. However, in the early phase of an accident it is unlikely that cost would be considered at all. For that reason we have set $x_1 = 0$.

A different method would be to construct two hypothetical accident scenarios. One could then adjust the probability that an accident occurs in one scenario until the decision makers feel that the consequences are comparable, i.e. that they are indifferent between the two consequences. For instance, one might imagine that there is a probability $p$ of an imminent release into the containment building which could be vented through filters immediately, exposing the local population to a noble gas release. The population would be advised to shelter. Alternatively, the risk could be taken to let the pressure build up while the population was evacuated. However, if the pressure ruptured the containment building, the release would not be filtered and the evacuating population would receive a significant dose. Suppose that a population of 400 were at risk. Suppose further that the dose that an individual would receive if exposed to an unfiltered release is 2 mSv, whereas the dose resulting from sheltering during a controlled filter release is 0.5 mSv. Then the choice would be between the following options:

1. *Controlled venting.* If there is a release, the population of 400 each have 1.5 mSv of their potential dose averted, assuming that they all shelter and no evacuation takes place. This gives a utility of:

$$
\begin{aligned}
u_{\text{vent}} = p &\times [1.0 \times 1.5 \times 400 + 1.0 \times 0.5 \times 400 + 100 \times 0.5 \\
&+ 0.0 \times (20 + 0.005 \times 400)] + (1 - p) \times a_2 \\
&\times [1.0 \times 0.8 \times 400 + 100 \times 0.4 \\
&+ 0.1 \times (20 + 0.005 \times 400)] \\
= p &\times 850 + (1 - p) \times a_2 \times 362.2
\end{aligned}
$$

2. *Evacuate while pressure builds up.* Assume that there is an 80 per cent probability that in the event of a release into the containment building all 400 are evacuated before the containment fails, hence averting all dose:

$$u_{\text{evac}} = p \times 0.8 \times [1.0 \times 2.0 \times 400 + 1.0 \times 0.5 \times 400 + 100 \times 0.5$$
$$+ 0.0 \times (20 + 0.100 \times 400)] + p \times 0.2 \times [1.0 \times 0.0 \times 400$$
$$+ 1.0 \times 0.8 \times 400 + 100 \times 1.0 + 0.0 \times (20 + 0.100 \times 400)]$$
$$+ (1 - p) \times a_2 \times [1.0 \times 0.05 \times 400 + 100 \times 0.0$$
$$+ 0.1 \times (20 + 0.100 \times 400)]$$
$$= p \times 924 + (1 - p) \times a_2 \times 26$$

For small $p$, i.e. a small chance of an accident, $u_{\text{vent}}$ is the larger, as it should be. The decision makers would not evacuate unless the risk became greater. One may ask them how large the risk of the accident would have to be before they would evacuate instead of waiting and venting/sheltering if the accident occurred. If they were to suggest that the probability $p$ would need to be about 12 per cent before they would evacuate, this would suggest that $a_2$ would be about 0.1. This is the value that we have taken in the example.

### 6.5.4 Analysis

Table 6.6 gives the result for each of areas 8, 9 and 10 of running the model with different probabilities of release. It can be seen that the model is

**Table 6.6** Countermeasure suggested by the model as a function of the probability of an accident

| Area | Probability of release | | | | | |
| | 0.001 | 0.05 | 0.1 | 0.2 | 0.5 | 1.0 |
|---|---|---|---|---|---|---|
| Area 8 | None | Shelter | Shelter | Evacuate | Evacuate | Evacuate |
| Area 9 | None | Shelter | Shelter | Shelter | Evacuate | Evacuate |
| Area 10 | Warn | Warn | Warn | Evacuate | Evacuate | Evacuate |

behaving reasonably sensibly. It only recommends evacuation when the probability of release is sufficiently high. Moreover, the probability of release has to be higher before evacuation of the large town in area 9 is recommended than before the evacuation of the less densely populated areas 8 and 10 is recommended.

A rather poorer feature of the model is that, when the probability is near zero, the model warns the population of area 10 but takes no action in areas 8 and 9. This is an effect of setting the balance between $y_1$ and $y_2$ to be

1 : 100 and then treating the decision in areas 8, 9 and 10 independently of each other. In treating area 10 independently, the model ignores the stress effects on the large population in area 9 and assumes that one can warn one area and not warn other areas. In practice, any warning will necessarily reach all areas. The use of constraint management techniques, to which reference has already been made, will avoid such unwanted effects within the models used within RODOS.

Investigating the model with sensitivity analysis techniques built into DPL has produced no other significant anomalies. However, this should not detract from the fact that this is an entirely hypothetical study. There is no intention to suggest that the judgemental values used here are 'correct' in any way. We simply wished to investigate the form of a model which may overcome the difficulties we have encountered in modelling the decisions faced when a nuclear accident threatens.

## 6.6 Discussion

It is possible that other decision analysts with more ingenuity than ourselves could model these problems without resort to event conditional attribute modelling. By careful definition of the attributes, it may be possible to deal with the issues we have resolved via distinct attribute hierarchies. However, we point to the following analogy. Act conditional subjective expected utility theories are formally equivalent to unconditional theories (Krantz *et al.*, 1971, Ch. 8); yet generally decision makers are more comfortable with conditional theories. Decision analyses are usually conducted in extensive rather than normal form, i.e. via decision trees and influence diagrams rather than decision tables (French, 1986). Thus we expect that here also decision makers will find it easier cognitively to use event conditional rather than unconditional attribute modelling. This is one of the issues we shall be investigating in the elicitation exercises referred to earlier.

It is also interesting to note a connection, albeit a tenuous one, with weighted expected utility models (Fishburn, 1988; French and Xie, 1994). Continuing the notation of Section 6.4: if $p$ is the probability of an accidental release and this is associated with the consequence set $C_1$, the various countermeasure strategies are to be ranked by their expected utilities:

$$pu_1(c_1) + (1-p)\alpha_2 u_2(c_2)$$

Since $p$ is constant for all countermeasure strategies, this takes the form of Chew's weighted expected utility model.[9] This example of a generalized expected utility theory has been proposed to explain departures from the standard expected utility model observed in the actual choice behaviour of decision makers. Perhaps this is a case of a generalized utility model arising through the application of 'standard' SEU to special contexts (French and Xie, 1994; see Bordley and Hazen, 1992, for another example).

---

[9] When $C$ is partitioned into $r$ subsets, the same observation can be made.

The above expression points to a further difficulty: $\alpha_2$ and $p$ enter the model together as a product. The elicitation of $\alpha_2$ may be confounded with the decision makers' discomfort in dealing with the probability of an accident. In a sense, it is difficulties such as these that have led to the development of descriptive models such as Chew's. One might wonder, therefore, whether the event conditional attribute modelling suggested here will be practicable. That is a question that only further work with real decision makers can answer.

# References

ADA Decision Systems 1996: *Decision Programming Language DPL.* Version 3.20. 2710 Sand Hills Road, Menlo Park, CA 94025.

Bordley, R.F. and Hazen, G. 1992: Non-linear utility models arising from unmodelled small world intercorrelations. *Management Science* **38**, 1010–17.

Ehrhardt, J., Päsler-Sauer, J., Schüle, O., Benz, G., Rafat, M. and Richter, J. 1993: Development of RODOS, a comprehensive decision support system for nuclear emergencies in Europe – an overview. *Radiation Protection Dosimetry* **50**, 195.

Fishburn, P.C. 1988: *Nonlinear preference and utility theory.* Brighton: Harvester Wheatsheaf.

French, S. 1986: *Decision theory: an introduction to the mathematics of rationality.* Chichester: Ellis Horwood.

French, S. 1996a: Multi-attribute decision analysis in the event of a nuclear accident. *Journal of Multi-Criteria Decision Analysis* **5**, 39–57.

French, S. 1996b: The framing of statistical decision theory: a decision analytic view. In Berger, J., Bernardo, J.M., Dawid, A.P. and Smith, A.F.M. (eds) *Bayesian Statistics 5: Proceedings of the 5th Valencia International Meeting on Bayesian Statistics.* Oxford: Oxford University Press, 147–64.

French, S. 1997: *Statistical decision theory. Kendall's advanced theory of statistics library.* London: Edward Arnold.

French, S., Halls, E. and Ranyard, D.C. 1995a: *Equity and MCDA in the event of a nuclear accident.* School of Computer Studies, University of Leeds. (A shorter version of this paper was given at the XII[th] International Conference on Multiple Criteria Decision Making in Hagen, 19–23 June 1995. RODOS(B)-RP(95)03.)

French, S., Ranyard, D. and Smith, J.Q. 1995b: *Uncertainty in RODOS.* Research Report 95.10, School of Computer Studies, University of Leeds. (Available by connecting to WWW at file://agora.leeds.ac.uk/scs/doc/reports/1995 or by anonymous ftp of the file scs/doc/reports/1995/95_10.ps.Z from agora.leeds.ac.uk. RODOS(B)-RP(94)05.)

French, S. and Vassiloglou, M. 1983: Examinations and multi-attribute utility theory. In French, S., Hartley, R., Thomas, L.C. and White, D.J. (eds) *Multi-objective decision making.* London: Academic Press, 307–22.

French, S. and Xie, Z. 1994: A review of recent developments in utility theory. In Sixtos Rios (ed.) *Decision making: trends and challenges.* Dordrecht: Kluwer, 15–31.

Janis, I.L. and Mann, L. 1977: *Decision making: a psychological analysis of conflict.* New York: Free Press.

Keeney, R.L. 1992: *Value-focused thinking: a path to creative decision making.* Cambridge, MA: Harvard University Press.

Keeney, R.L. and Raiffa, H. 1976: *Decisions with multiple objectives: preferences and value trade-offs.* New York: John Wiley and Sons.

Kelly, G.N. 1994: Private communication.

Kelly, G.N. and Ehrhardt, J. 1995: RODOS – a comprehensive decision support system for off-site emergency management. *Proceedings of the Fifth Topical Meeting on Emergency Preparedness and Response.* American Nuclear Society.

Krantz, D.H., Luce, R.D., Suppes, P. and Tversky, A. 1971: *Foundations of measurement*, Vol. 1. New York: Academic Press.

Lochard, J., Schneider, T. and French, S. 1992: *International Chernobyl Project – input from the Commission of the European Communities to the evaluation of the relocation policy adopted by the former Soviet Union.* CEC, Luxembourg, EUR 14543 EN.

Phillips, L.D. 1984: A theory of requisite decision models. *Acta Psychologica* **56**, 29–48.

Phillips, L.D. 1995: Private communication.

Richard, R., van der Pligt, J. and de Vries, N. 1996: Anticipated regret and time perspective: changing sexual risk behaviour. *Journal of Behavioural Decision Making* (in press).

Russo, J.E. and Shoemaker, P.J. 1989: *Decision traps: the ten barriers to brilliant decision making and how to overcome them.* New York: Doubleday.

Savage, L.J. 1954: *The foundations of statistics.* New York: John Wiley and Sons.

Walmod-Larsen, O. (ed.) 1995: *Intervention principles and levels in the event of a nuclear accident.* TermaNord 1995:507, Risø National Laboratory, DK4000 Roskilde, Denmark.

# 7 Optimal decisions that reduce flood damage along the Meuse: an uncertainty analysis[1]

J M van Noortwijk, M Kok,
R M Cooke

## 7.1 Introduction

Around Christmas 1993, the Dutch river Meuse flooded due to an extreme discharge at Borgharen (near the Dutch–Belgian border) of 3120 m$^3$/s (see Fig. 7.1). Discharges at Borgharen larger than 3120 m$^3$/s have a probability of occurrence of about once in 150 years. In Limburg, the flood caused damage of about 250 million Dutch guilders, a flooded area of about 18 000 hectares, an evacuation of about 8000 people, and therefore raised emotions.

To investigate and compare decisions that reduce future losses due to flooding, the Dutch Minister of Transport, Public Works and Water Management initiated the project 'Investigation of the Meuse flood'. The project was carried out in the period March–November 1994 by about 90 researchers, mainly from Delft Hydraulics and the Dutch Ministry of Transport, Public Works and Water Management (Rijkswaterstaat), and had a budget of six million Dutch guilders. The study was supervised by a committee, named after its chairman Dr B.C. Boertien, which had the following members: two from the local waterboards, three from the province of Limburg, two from the Rijkswaterstaat, one mayor, and one representative from Belgium.

Roughly, the Dutch Meuse can be subdivided into two parts: (1) the upstream area in Limburg without dykes, but with small embankments that were all overtopped during the 1993 flood; and (2) the downstream area with dykes that were high and strong enough to prevent the protected

[1] The authors acknowledge comments from Tim Bedford.

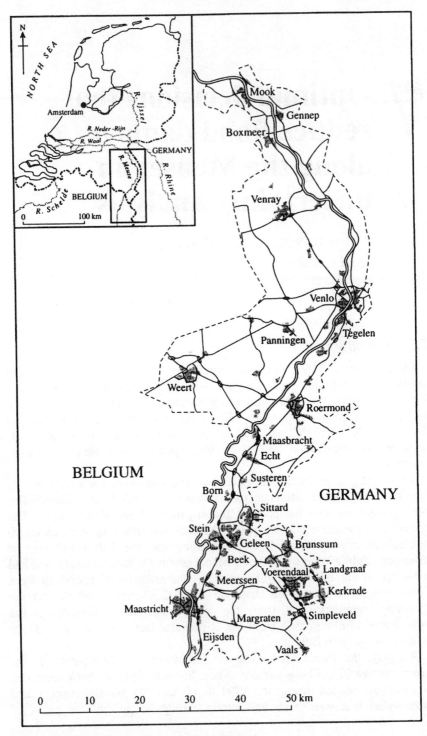

**Fig. 7.1** The Meuse river in Limburg, The Netherlands

polders from flooding. The subject of study is the upstream area of the
Meuse in Limburg. The results[2] of the project are reported in one main
report (Delft Hydraulics, 1994a) and 14 subreports; this paper summarizes
Subreport 14 on the uncertainty analysis (Delft Hydraulics and Delft
University of Technology, 1994: hereafter DHDUT, 1994). For a summary
of the main report, see Kok (1995).

Since decisions that reduce flood damage must be made under uncertainty, Section 7.2 presents three types of criteria to compare decisions in
uncertainty: the criterion of minimal expected loss, the criterion of minimal
uncertainty in the loss, and the criterion of maximal safety. For flooding of
the river Meuse, the most important uncertainties are the river discharge at
Borgharen (Section 7.3.1), the flood damage given the discharge (Section
7.3.2), the downstream water levels along the Meuse given the discharge
(Section 7.3.3), and the costs and yields of decisions (Section 7.3.4).

After representing the above uncertainties with probability distributions,
the uncertainty in the loss due to flooding remains to be determined. For
flooding of the Meuse, the loss is defined as the net present discounted value
of the costs of decisions minus the yields of decisions plus the mean flood
damage over an unbounded time-horizon (Section 7.3.5). Section 7.4 reports
the results of an uncertainty analysis in which the uncertainties are determined in the mean flood damage and in the loss, for the present situation and
for five strategies (combinations of decisions). In Section 7.5, we discuss
which strategy is optimal according to each decision criterion. Some
definitions of probability distributions can be found in the appendix (p. 172).

## 7.2   Decision making under uncertainty

### 7.2.1   Optimal decisions

Optimal decisions that reduce flood damage must be made under uncertainty
and can be obtained using decision theory. Following the treatments of
DeGroot (1970) and Savage (1972), a decision problem is a problem in
which the decision maker has to choose a decision $d$ (or a combination of
decisions) from the set of all possible decisions $\mathcal{D}$, where the consequences
of decision $d$ depend on the unknown value $w$ of the state of the world $W$
(for example, the discharge of the Meuse at Borgharen). Optimal decisions
can be defined with respect to the following three decision criteria: the
criterion of minimal expected loss, the criterion of minimal uncertainty in
the loss and the criterion of  maximal safety.

*The set of possible decisions*   Five strategies that reduce future losses due
to flooding have been selected for further investigation in Delft Hydraulics

---

[2] At the beginning of 1995, the high water levels on the Dutch rivers were world news. Also,
the Meuse flooded due to a maximum discharge at Borgharen of 2870 $m^3/s$. The 1995 data,
however, are not incorporated into the present uncertainty analysis.

(1994a, h). These strategies were developed during a careful screening process of all measures that might result in a reduction of flood damage and in new natural development of the Meuse. Also, measures proposed by society were considered, but none of them were attainable. Finally, a limited number of measures were selected and combined in strategies. Roughly, the strategies can be subdivided into three categories (see Table 7.1) on the

**Table 7.1** The investigated strategies with the measures included (+)

| Measure | Strategy | | | | |
|---|---|---|---|---|---|
| | 1 | 2a | 2b | 2c | 3 |
| Lowering the summer bed | | | | | |
| North of Limburg | + | + | + | – | – |
| Middle of Limburg | + | + | + | – | – |
| South of Limburg | + | – | – | – | – |
| Lowering the winter bed: north of Limburg | – | + | – | – | – |
| Widening the summer bed: south of Limburg | – | + | + | + | – |
| Constructing embankments and dykes (km) | 55 | 58 | 62 | 128 | 137 |

basis of lowering the summer bed (strategy 1); natural development of the Meuse in the south of Limburg (strategy 2abc), and the construction of embankments and dykes only (strategy 3). Note that all strategies cover the construction of embankments and dykes around the remaining bottlenecks along the Meuse.

*Decisions with minimal expected loss*    Let $L(w, d)$ be the loss when the decision maker chooses decision $d$ and when the value of $W$ is $w$. In flooding, the loss function equals the costs of decision $d$, say $c(w, d)$, minus the yields of decision $d$, say $y(w, d)$, plus the remaining flood damage, say $s(w, d)$. Hereby, $c(w, d)$ represents the costs of extracting sand and gravel from the river bed (to lower or widen it) and of constructing embankments and dykes; $y(w, d)$ represents the yields of extracting sand and gravel. Hence, the *loss* can be written as

$$L(w, d) = c(w, d) - y(w, d) + s(w, d) \qquad (7.1)$$

with $c(w, \varnothing) = y(w, \varnothing) = 0$ (no decisions are made). For any decision $d \in \mathcal{D}$, the expected loss is given by $E(L(W, d))$. The decision maker can best choose, if possible, the decision $d^*$ whose expected loss is minimal. A decision $d^*$ is called an *optimal decision* when $E(L(W, d^*)) = \min_{d \in \mathcal{D}} E(L(W, d))$.

*Decisions with minimal uncertainty in the loss*    Uncertainty in the loss can be important when a decision maker has to choose between two decisions with equal expected loss. A possible decision rule could be to choose the decision with minimal uncertainty in the loss.

*Decisions with maximal safety* Instead of minimizing the loss, one might prefer maximizing the safety (or the utility). Some people living along the Meuse were interviewed about their subjective feelings of safety under different flooding conditions (see Delft Hydraulics, 1994b). Since minimizing loss and maximizing safety cannot both be achieved, a possible decision rule is choosing the decision for which an acceptable safety level will yet be attained.

### 7.2.2 The expected loss of a decision

In general, decision problems are based on the consequences of the uncertain state of the world $W$. For flooding of the Meuse, the state of the world is characterized by the following five random quantities: the river discharge at Borgharen, the downstream water levels along the Meuse given the Borgharen discharge, the flood damage given the water level, the costs of decisions and the yields of decisions. Although more uncertainties can be identified, these five are the most relevant, for small uncertainties pale into insignificance beside large uncertainties.

The main aim of this paper is to compute the probability distribution of the loss function, eq. (7.1), and its expected value. We determine the joint probability density function of the above five random quantities by formulating the decision problem in terms of an influence diagram. For a brief introduction to influence diagrams, we refer to Barlow and Pereira (1993) and Jae and Apostolakis (1992).

To obtain an influence diagram for flooding of the Meuse, we split the joint probability density function up into conditional probability distributions that can be easily assessed (see Fig. 7.2). The main source of uncertainty in the event of a Meuse flood is the maximal river discharge $Q$ at Borgharen in $m^3/s$ (in Fig. 7.2, $Q$ is displayed as a chance node). To avoid calculational burden, we discretize the probability distribution of the discharge $Q$ into the intervals $(q_{i-1}, q_i], i = 1, \ldots, 9$ (see Table 7.2). Given a

**Table 7.2** List of discharges at Borgharen ($m^3/s$) according to which the probability distribution of the discharge has been discretized

| | | | | List of discharges ($m^3/s$) | | | | | |
|---|---|---|---|---|---|---|---|---|---|
| $i$ | 0 | 1 | 2 | 3 | 4 | 5 | 6 | 7 | 8 | 9 |
| $q_i$ | 2000 | 2120 | 2500 | 2750 | 2990 | 3120 | 3305 | 3545 | 3860 | $\infty$ |

particular discharge at Borgharen, we can obtain the downstream water level with the one-dimensional physical model ZWENDL developed by Rijkswaterstaat. The water level at a given location mainly depends on the discharge and the river geometry. Since we can measure the river geometry, we may regard it as a known quantity resulting in the flood depth (a deterministic

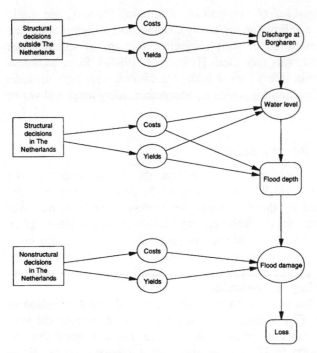

**Fig. 7.2** Influence diagram of decisions that reduce flood damage along the Meuse river

node). Given the water level and the geometry, the damage assessment model (see Delft Hydraulics, 1994e; De Jonge *et al.*, 1996) determines whether, and to what extent, immovables will be damaged due to flooding. For a number of flood depths, this model estimates the number of flooded houses, industries, farms, greenhouses (hectares), roads (km) and government agencies.

To reduce damage due to flooding, we can identify three types of decisions (decision nodes in Fig. 7.2): first, structural decisions to be taken upstream in Germany, Belgium and France, which affect the Borgharen discharges; second, structural decisions to be taken downstream in the Netherlands which affect the water levels via changes in the geometry and in the hydraulic roughness; third, non-structural decisions which affect the flood damage via heightening of meter cupboards, making the furniture water-resistant and improving the water level predictions. Given the costs of decisions, the yields of decisions, and the flood damage, we can now obtain the loss.

## 7.3    Modelling uncertainty

In modelling the main uncertainties in our decision problem, being the discharge, the flood damage, the water level, the costs, the yields and the loss, we use the arguments of Barlow (1995): (1) in the presence of a large amount of data, summarization is useful; (2) a decision-theoretic tool should consist of a small number of parameters; (3) possible prior information should be used for analysing data; and (4) uncertainty can best be represented by probability distributions.

### 7.3.1 Uncertainty in the discharge

To derive a probabilistic model for the Borgharen discharge, we extend the frequentist approach of Delft Hydraulics (1994c) with uncertainty distributions. In essence, the frequentist approach has resulted in a probability of exceedance of a discharge $q$ of the form

$$\Pr\{Q > q \mid \phi, \theta\} = \Pr\{Q > q_0 \mid \phi\}\Pr\{Q > q \mid Q > q_0, \theta\}$$
$$= \phi \exp\{-(q - q_0)/\theta\} \qquad (7.2)$$

for $q > q_0$ and $\phi$, $\theta > 0$. Rather than treating $\phi$ and $\theta$ as known parameters, an uncertainty analysis requires regarding them as unknown random quantities. The choice for the threshold $q_0$ has been motivated by the decision problem: $q_0$ is the largest discharge for which the Meuse does not exceed the summer bed and for which no flood damage occurs, i.e. $q_0 = 2000 \ \mathrm{m^3/s}$ (see Delft Hydraulics, 1994e).

#### 7.3.1.1 *The probability that flood damage occurs*

For determining the probability of flood damage, i.e. of the event $Q > q_0$, we need to know the relative frequency of years in which flood damage occurs in a potentially infinite sequence of hydrological years. Suppose that the order in which the floods occur is irrelevant, or, in other words, the years are exchangeable. Furthermore, we define the random quantity $V_i$ as follows: $V_i = 1$, if flood damage occurs in year $i$, and $V_i = 0$, if flood damage does not occur in year $i$, where $i \in \mathbb{N}$. By de Finetti's representation theorem (1937), there exists a unique probability distribution $P$ such that the joint probability density function of $V_1, \ldots, V_n$ can be written as a mixture of conditionally independent Bernoulli trials:

$$p(v_1, \ldots, v_n) = \int_0^1 \prod_{i=1}^n \phi^{v_i}(1 - \phi)^{1 - v_i} \, dP(\phi) = \int_0^1 \prod_{i=1}^n l(v_i \mid \phi) \, dP(\phi) \qquad (7.3)$$

where $l(v_i \mid \phi)$ is the likelihood function of the observation $v_i$. The random quantity $\Phi$ may be interpreted as the limiting relative frequency of discharges larger than $q_0$, i.e. $\lim_{n \to \infty}[(\sum_{i=1}^n V_i)/n]$, and the probability distribution $P$ as the representation of beliefs about $\Phi$ (for details, see Cooke, 1991, Ch. 7; Bernardo and Smith, 1994, Ch. 4).

As soon as data become available, the prior distribution $P$ on $\Phi$ can be updated to the posterior distribution by Bayes' theorem. For an uncertainty analysis, it is convenient when the prior distribution enables the posterior distribution to be expressed in explicit form. It is well known that the beta distribution $Be(\phi \,|\, a, b)$ has this property. With a Bernoulli likelihood function, both the prior distribution and the posterior distribution are a beta distribution: i.e. $Be(\phi \,|\, a, b)$ and $Be(\phi \,|\, a + \sum_{i=1}^{n} v_i, b + n - \sum_{i=1}^{n} v_i)$, respectively. The beta distribution is said to be conjugate with respect to the Bernoulli likelihood function (DeGroot, 1970, Ch. 9). Note that the (prior) predictive limiting relative frequency of discharges larger than $q_0$ is given by

$$\Pr\{Q > q_0\} = \int_0^1 \phi \, Be(\phi \,|\, a, b) \, d\phi = a/(a + b) \tag{7.4}$$

### 7.3.1.2  The probability of exceedance of a discharge

The last step in determining the probability distribution of the discharge $Q$ is to obtain the conditional probability $\Pr\{Q > q \,|\, Q > q_0\}$. Define the random quantity $Q_j$ to be the maximal Borgharen discharge in hydrological year $j$, where $j \in \mathbb{N}$. If the discharge causes flood damage, i.e. if $Q_j > q_0$, let $X_j = Q_j - q_0 \in \mathbb{R}_+$ for $j \in \mathbb{N}$ where $\mathbb{R}_+ = [0, \infty)$. Furthermore, we assume that the joint probability density function of $X_1, \ldots, X_m$ can be written as a function of the statistic $\sum_{j=1}^{m} X_j$. Then, with Mendel (1989) and Misiewicz and Cooke (1997), we have a mixture of exponentials:

$$p(x_1, \ldots, x_m) = \int_0^\infty \prod_{j=1}^{m} \frac{1}{\theta} \exp\left\{-\frac{x_j}{\theta}\right\} dP(\theta) = f_m\left(\sum_{j=1}^{m} x_j\right) \tag{7.5}$$

where the random quantity $\Theta$ may be interpreted as the limiting average discharge larger than $q_0$, i.e. $\lim_{m \to \infty} [(\sum_{j=1}^{m} X_j)/m]$ (provided $E(X_j) < \infty$ for $j \in \mathbb{N}$). The probability distribution $P$ represents the uncertainty in $\Theta$. If the joint probability density function of $X_1, \ldots, X_m$ satisfies eq. (7.5), then $X_1, \ldots, X_m$ are called $l_1$-isotropic.

The probability distribution that is conjugate with respect to the exponential likelihood function is the inverted gamma distribution $Ig(\theta \,|\, v, \mu)$. By updating this prior distribution with the observations $x_1, \ldots, x_m$, the posterior distribution is the inverted gamma distribution $Ig(\theta \,|\, v + m, \mu + \sum_{j=1}^{m} x_j)$. The (prior) predictive conditional probability that $Q > q$, given $Q > q_0$, is called the gamma-gamma distribution (see e.g. Bernardo and Smith, 1994, Ch. 3):

$$\Pr\{Q > q \,|\, Q > q_0\} = \int_0^\infty \exp[-(q - q_0)/\theta] \, Ig(\theta \,|\, v, \mu) d\theta = \left[\frac{\mu}{\mu + q - q_0}\right]^v \tag{7.6}$$

for $q > q_0$.

Let us assume the random quantities $\Phi$ and $\Theta$, representing the uncertainties in the limiting relative frequency of discharges larger than $q_0$ and the limiting average discharge larger than $q_0$, respectively, to be independent. In conclusion, we can easily obtain the (prior) predictive probability of exceedance of a discharge $q$ by using eqs (7.4) and (7.6):

$$\Pr\{Q > q\} = \int_0^\infty \int_0^1 \phi \exp\left[-(q - q_0)/\theta\right] \operatorname{Be}(\phi \,|\, a, b) \operatorname{Ig}(\theta \,|\, v, \mu)\, d\phi\, d\theta \quad (7.7)$$

for $q > q_0$. As was to be required, eq. (7.7) extends eq. (7.2) with the uncertainty distributions $\operatorname{Be}(\phi \,|\, a, b)$ and $\operatorname{Ig}(\theta \,|\, v, \mu)$. Actually, we have approximated the tail of the probability distribution of the discharge with a mixture of exponentials. In the presence of $m$ observations which are larger than $q_0$, i.e. for $x_j = q_j - q_0$, $j = 1, \ldots, m$, the parameters $a$, $b$, $\mu$ and $v$ must be replaced by $a + m$, $b + n - m$, $\mu + \sum_{j=1}^m x_j$ and $v + m$, respectively, where $m = \sum_{i=1}^n v_i$.

### 7.3.1.3  *Prior information and observed discharges*

Next, we shall determine the parameters of the two uncertainty distributions that we have chosen. For this purpose, we consider two types of distributions: a non-informative and an informative prior distribution.

*Non-informative prior distribution*  There is no prior information to be taken into account, i.e. the prior uncertainty is very large. To express 'very large uncertainty' in probabilistic terms, we can use several non-informative prior distributions (see Berger, 1985, Ch. 3; Bernardo and Smith, 1994, Ch. 5). In our opinion, we can best use a non-informative prior distribution whose posterior mean equals the sample mean. Under this condition, the non-informative prior densities for $\phi$ and $\theta$ are $\phi^{-1}(1 - \phi)^{-1} I_{[0,1]}(\phi)$ and $\theta^{-2} I_{(0,\infty)}(\theta)$, respectively.

*Informative prior distribution*  There is prior information to be taken into account: the data on floods that occurred between 1400 and 1910. These data have been interpreted from Gottschalk (1975, 1977) and the KNMI (see DHDUT, 1994). Although the real historical discharges are unknown and the river geometry is changed, these references do mention whether there was flood damage or a catastrophe attended with drowned people, dyke bursts, flooded polders, carried away houses and collapsed bridges. Probably, the most serious flood of the river Meuse, up to now, is the flood of 1643. On the basis of the amount of flood damage, we assume for the Borgharen discharge $q$ that (using Table 7.2):

- $q \le q_0$, if the references do not mention any flood damage;

- $q_0 < q \le q_2$, if the references mention flood damage, but no catastrophe;

- $q > q_2$, if the references mention a catastrophe.

From the historical data of the period 1400–1910, we have derived the following estimates: $\Pr\{Q > q_0\} = 0.1373$ and $\Pr\{Q > q_2\} = 0.0275$. Hence, it follows that $\Pr\{Q > q_2 \,|\, Q > q_0\} = 0.2003$. Since the historical data are probably less reliable for determining $\Pr\{Q > q_0\}$ than for determining

$\Pr\{Q > q_2 \mid Q > q_0\}$, and since the weight of the historical data in comparison with the data of the years 1911–1993 should not be too large, we assume that the prior information of the years 1400–1910:

- may not be used for determining the probability distribution of the probability that the discharge is larger than $q_0$ (i.e. of $\Phi$); hence, let the prior density of $\Phi$ be non-informative: $\phi^{-1}(1 - \phi)^{-1}I_{[0,1]}(\phi)$;

- may be used as five imaginary data for determining the prior distribution of the average discharge larger than $q_0$ (i.e. of $\Theta$); hence, let the prior density of $\Theta$ be informative: $\mathrm{Ig}(\theta \mid 5, \mu)$ where $\mu$ follows from eq. (7.6) if $q = q_2$.

In turn, Bayes' theorem can be applied to update the historical prior information with the observations of the years 1911–1993. These 83 observations are displayed in Fig. 7.3 and can be found in DHDUT, (1994): 13 observations are larger than $q_0$ and three are larger than $q_2$. The posterior

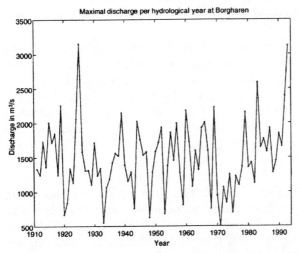

**Fig. 7.3** Maximal discharge per hydrological year at Borgharen during 1911–1993

distribution of $\Phi$, the limiting relative frequency of discharges larger than $q_0$, given the observations is the beta distribution $\mathrm{Be}(\phi \mid 13, 70)$. The posterior distribution of the limiting average discharge $\Theta$ is the inverted gamma distribution $\mathrm{Ig}(\theta \mid 18\ 5630.12)$. The (posterior) predictive probability of exceedance of a discharge $q$, eq. (7.7), is shown in Fig. 7.4. It slightly differs from the frequentist result (Delft Hydraulics, 1994c) in the sense that the probabilities of exceedance of very extreme discharges are slightly larger, just because the uncertainty is taken into account (see DHDUT, 1994).

### 7.3.2  Uncertainty in the flood damage

Beside the uncertainty in the occurrence frequency of the Borgharen discharges, the uncertainty in the flood damage is also important. Since the

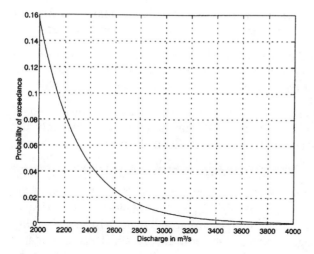

**Fig. 7.4** Posterior predictive probability of exceeding the Borgharen discharge $q$

Meuse flood in December 1993 caused much damage, there is a large volume of data available (at least on damage caused by flood depths up to about 1 m). For example, about 5600 houses were flooded. For about 4600 houses, damage data are known with an average damage of about 15 600 Dfl per house (see Table 7.3). The damage assessment was done by experts

**Table 7.3** The average flood damage per house (Dutch guilders) and the number of flooded houses against the flood depth (cm) due to the Meuse flood in December 1993

| Flood depth (cm) | 0–10 | 10–17.5 | 17.5–25 | 25–50 | 50–75 | >75 |
|---|---|---|---|---|---|---|
| Average flood damage (Dfl) | 7500 | 10 500 | 12 200 | 15 200 | 18 600 | 18 800 |
| Number of flooded houses | 249 | 578 | 374 | 1262 | 1052 | 1059 |

from the insurance companies by order of the Dutch government, because the government partly covered the flood damage. The methodology to determine the uncertainty in the flood damage will be explained in the light of the damage category 'houses'.

To deal with a large amount of observations (e.g. 4600 flooded houses), it is necessary to summarize them. As we have argued in Section 7.2.2, the probability distributions of the flood damage can best be assessed conditional on disjunct classes of flood depths (see Table 7.3). The conditional probability distribution of the flood damage, given a flood depth class, must satisfy the following three criteria:

- flood damage is non-negative (the assumption that the damage has a normal distribution can cause inconsistencies, especially if we are interested in the tails);

- per flood depth, a proper summarizing statistic is the number of flooded houses and the corresponding sum of flood damages to these houses;

- the predictive flood damage equals the outcome of the damage assessment model MAAS-GIS in which the Meuse has been subdivided into 200 000 'pieces' (see Delft Hydraulics, 1994e; De Jonge *et al.*, 1996).

Let the random quantity $Z_i \geq 0$ be the flood damage to house $i$ for a particular flood depth class, where $i \in \mathbb{N}$. For the decision maker only the number of flood damages and the sum of flood damages are important. Therefore, we may assume that the decision maker is indifferent to the way this sum is composed. In other words, we may assume that the joint probability density function of $Z_1, \ldots, Z_n$ can be written as a function of the statistic $\sum_{i=1}^{n} Z_i$, i.e. $p(z_1, \ldots, z_n) = f_n(\sum_{i=1}^{n} z_i)$ (see Mendel, 1989). Hence, per flood depth class, $Z_1, \ldots, Z_n$ are $l_1$-isotropic and there exists a probability distribution of the average flood damage per house, $\Lambda$, which can be updated with the observations in Table 7.3 using Bayes' theorem. Analogous to the discharges, we use the inverted gamma distribution $\text{Ig}(\lambda \mid v, \mu)$ to represent the uncertainty in the average flood damage per house $\Lambda$. As a consequence, the sum of flood damages to $n$ houses, $U_n = \sum_{i=1}^{n} Z_i$, has a gamma-gamma distribution:

$$\text{Gg}(u_n \mid v, \mu, n) = \int_0^{\infty} \text{Ga}(u_n \mid n, 1/\lambda) \text{Ig}(\lambda \mid v, \mu) \, d\lambda \qquad (7.8)$$

with mean $E(U_n) = n[\mu/(v-1)]$ (see Bernardo and Smith, 1994, Ch. 3). Because the expected flood damage to $n$ houses has to be equal to the outcome of the damage assessment model, we choose the non-informative prior density $\lambda^{-2} I_{(0,\infty)}(\lambda)$. In the presence of the observations $z_1, \ldots, z_m$, this results in the inverted gamma distribution $\text{Ig}(\lambda \mid m+1, \sum_{i=1}^{m} z_i)$. Indeed, $E(U_n) = n[(\sum_{i=1}^{m} z_i)/m]$. Note that the larger the predictive number of houses, the larger the absolute uncertainty, but the smaller the relative uncertainty.

Given the flood depth class, we can best summarize the observations by the number of houses, industries, farms, greenhouses (hectares), roads (km), government agencies, and the corresponding sum of flood damages to these objects. The unit of area for greenhouses (hectare) and the unit of length for roads (km) are chosen in such a way that the uncertainties per damage category are of the same order of magnitude (for the underlying mathematics, see Van Noortwijk *et al.*, 1994).

Up to now, we have considered the ideal case, i.e. the case for which the probability distribution of the flood damage can be determined for every flood depth class separately. Eventually, the number of flood depth classes for other categories than houses appeared to be so large (e.g. for industries: 60) that it was hardly possible to assess the number of flooded objects for every flood depth class separately. With the damage assessment model, we can only determine the number of flooded objects and the corresponding flood damage aggregated per damage category. On the basis of the number of objects and the flood damage per category, and the detailed information on houses, we may approximate the probability distributions of the flood damage to other categories than houses. From DHDUT (1994), it appears that the flood damages summed over all flood depth classes can be approximated by the gamma-gamma distribution $Gg(\nu, \mu, n)$ for which:

$n$ = number of flooded objects from damage assessment model

$\nu = 0.5 \times$ number of flooded objects in December 1993

$\mu = [(\nu - 1)/n] \times$ total damage to flooded objects

The number of flooded objects in December 1993 determines the uncertainty in the average damage per object. The larger the amount of data of 1993, the smaller the uncertainty in the average damage. Finally, the probability distribution of the total damage summed over all damage categories can be approximated by a gamma distribution with equal first and second moments (see DHDUT, 1994). In Fig. 7.5, we present the expected flood damage summed over all categories for every discharge in Table 7.2, calculated by the damage assessment model for the present situation (see Delft Hydraulics, 1994e).

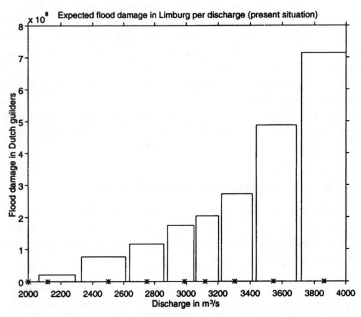

**Fig. 7.5** Expected flood damage in Limburg given the Borgharen discharges (\*) for the present situation

### 7.3.3   Uncertainty in the water level

Given a discharge at Borgharen, we can obtain the water level at locations downstream with the physical model ZWENDL (see Delft Hydraulics,

1994d). Beside the uncertainty in the discharge, the added uncertainty in the water level is also important. Conditional on the Borgharen discharge, the uncertainty in the water level is twofold:

- the uncertainty in *the shape of the waterwave* at Borgharen (see Delft Hydraulics, 1994d);

- the uncertainty in *the physical model* with which the water levels have been determined.

The two types of uncertainty are complementary: the larger the discharge at Borgharen, the smaller the variation in the shape of the water wave, but the larger the uncertainty in the physical model.

On the basis of the two types of uncertainty in the water levels, the uncertainty in the frequency of occurrence of the water levels has to be determined. It is well known that there exists an approximate power law between the discharge (e.g. at Borgharen) and the downstream water levels (see, e.g., Shaw, 1988, Ch. 6):

$$q - q_0 = a(h - h_0)^b, \quad q > q_0 > 0, \quad h > h_0 > 0 \tag{7.9}$$

with $q$ the upstream discharge ($m^3/s$) and $h$ the downstream water level (m).

Since the power law between discharge and water level is just an approximation, the quantities $a$ and $b$ are unknown. Hence, we are interested in the joint probability density function of the random quantities $A$ and $B$, where $A$ and $B$ may be dependent. For just this situation, Cooke (1994) has developed a technique to obtain the marginal distributions of $A$ and $B$, and their correlation, on the basis of the uncertainties in the (logarithm of the) water level given the (logarithm of the) discharge.

Although the transformation from discharges at Borgharen to downstream water levels introduces extra uncertainties, they are not so large that they should be taken into account (for details, see DHDUT, 1994).

### 7.3.4   Uncertainty in the costs and the yields of decisions

The probability distributions of the costs and the yields of decisions are based on Delft Hydraulics (1994a, f, g). They are all assumed to be gamma distributed with parameters that are assessed using informal expert judgment (for details, see DHDUT, 1994).

*Costs of extracting sand and gravel*   As the 5th and the 95th percentiles of the probability distribution of the costs of extracting sand and gravel in the north and the middle of Limburg, we use the mean in DHDUT (1994) plus or minus a deviation of 10 per cent. Since there is more knowledge about the price of extracting sand and gravel in the south of Limburg, this deviation from the mean is taken to be plus or minus 5 per cent.

*Yields of extracting sand and gravel*   Since the yields of extracting sand and gravel are just a matter of supply and demand, the uncertainty in the yields is probably larger than the uncertainty in the costs. Hence, we assume the 5th and the 95th percentiles to be the mean plus or minus a deviation of 17 per cent.

*Costs of constructing embankments and dykes*  For the construction of embankments and dykes, one usually expresses the uncertainty in the costs by an uncertainty factor $F$, where $F \approx 2.5$. Instead, we regard $F$ as a random quantity that satisfies the following two requirements:

$$\Pr\{F \le 2\} = 0.1 \cap \Pr\{F > 2.5\} = 0.95 \tag{7.10}$$

Next, we scale the gamma distribution satisfying these two equalities with the expected costs in DHDUT (1994).

*Dependencies between costs and yields of decisions*  Since the yields of sand and gravel do not depend so much on the locations of extraction, we assume the yields per region (north, middle and south) to be dependent with a rank correlation of 0.75. The costs of extracting sand and gravel are dependent, but less than the yields, since the costs are partly determined by the location. Hence, we assume a rank correlation between the costs per region of 0.5. Furthermore, the costs and yields both depend on the same price level: by assumption, a rank correlation of 0.2.

### 7.3.5  Uncertainty in the loss

On the basis of the probability distribution of the discharge, we can derive an approximate expression for the loss function in eq. (7.1): the costs of decisions minus the yields of decisions plus the remaining flood damage. Since we have to find an optimum balance between the initial investment costs and the future flood damage costs, we can best compare decisions over an unbounded time-horizon by using the discounted cost criterion (see Van Noortwijk and Peerbolte, 1995). By discounting the future costs, yields and flood damage, the loss can be approximated by (see Delft Hydraulics, 1994e):

$$L(w, d) = \sum_{j=0}^{k} \tilde{\alpha}^{j} [c_j(d) - y_j(d)]$$

$$+ \frac{\alpha}{1-\alpha} \sum_{i=1}^{9} \Pr\{q_{i-1} < Q \le q_i | \phi, \theta\} \times \frac{[s_{i-1}(d) + s_i(d)]}{2} \tag{7.11}$$

where $\Pr\{Q > q | \phi, \theta\} = \phi \exp[-(q - q_0)/\theta]$ and the vector $w = [\mathbf{c}(d), \mathbf{y}(d), \mathbf{s}(d), \phi, \theta]$ is uncertain. The loss function in eq. (7.11) represents the net present discounted value of the costs minus the yields plus the remaining mean flood damage over an unbounded horizon, where

$d$ = decision or strategy (Section 7.2.1)

$k$ = 15 years (time-horizon for carrying out decision $d$; DHDUT, 1994)

$c_j(d)$ = costs of decision $d$ in year $j$ (Section 7.3.4)

$y_j(d)$ = yields of decision $d$ in year $j$ (Section 7.3.4)

$q_i$ = discharge at Borgharen in m$^3$/s, $i = 0, 1, ..., 9$ (Table 7.2)

$\phi$ = limiting relative frequency of discharges larger than $q_0$ (Section 7.3.1)

$\theta$ = limiting average discharge larger than $q_0$ (Section 7.3.1)

$s_i(d)$ = total flood damage to houses, industries, farms, greenhouses, roads, and government agencies per discharge $q_i$ when making decision $d$, where $s_0(d) = 0$ and $s_9(d) = s_8(d)$, $i = 0, 1, ..., 9$ (Section 7.3.2)

$r$ = discount rate per year (5 per cent)

$g$ = growth rate of capital in the flood plain per year (1 per cent)

$\tilde{a} = [1 + (r/100)]^{-1}$

$a = [1 + ((r - g)/100)]^{-1}$

As has been discussed in Section 7.2, an optimal decision $d^*$ with respect to the probability distribution of the random vector $W$ is the decision whose expected net present discounted value of the costs minus the yields plus the remaining mean flood damage, over an unbounded time-horizon, is minimal. Since the discount rate $r$ and the growth rate of capital $g$ are essentially based on an agreement on comparing decisions over a long time-horizon, we assume these to be constant. The expected values $E(\Pr\{Q > q \mid \Phi, \Theta\})$ and $E(S_i(\emptyset))$, $i = 0, 1, ..., 8$, are shown in Figs 7.4 and 7.5, respectively. We have determined both the expected loss and the uncertainty in the loss by using Monte Carlo simulation.

## 7.4   The results of the uncertainty analysis

When the uncertainties in the discharge, the flood damage, and the costs and the yields of decisions are given, the uncertainty in the loss can be determined. We consider the present situation and the five selected strategies of Section 7.2.1. The results of the uncertainty analysis have been obtained by the Monte Carlo simulation program UNICORN (see Cooke, 1995). On the basis of the probability distributions of the random quantities $C_j(d)$, $Y_j(d)$, $j = 1, ..., k$, $S_i(d)$, $i = 1, ..., 9$, $\Phi$ and $\Theta$, and their correlations, the UNICORN program approximates the probability distribution of an analytic function of these random quantities, like the loss function, by performing dependent Monte Carlo sampling as described in Cooke and Waij (1986). The sample size was 200 000 for the present situation and 100 000 for the five strategies; the number of random quantities was 37 per strategy (for details, see DHDUT, 1994).

### 7.4.1    Uncertainty in the mean flood damage

In Figs 7.6 and 7.7, some probabilistic characteristics of the damage due to flooding of the Meuse are displayed.

In Fig. 7.6, the approximated probability density function is shown of $s(w, \varnothing)$, the present discounted value of the mean flood damage over an unbounded horizon when no decisions are taken. In Fig. 7.7, the results of

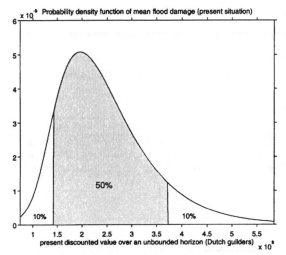

**Fig. 7.6** The probability density function, and its 10th and 95th percentiles, of the present discounted value of the mean flood damage over an unbounded horizon when no decisions are made

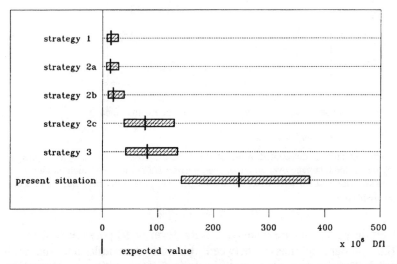

**Fig. 7.7** The present discounted value of the mean flood damage over an unbounded horizon (in millions of Dutch guilders). For the present situation and for five strategies, the 80 per cent probability bars and the expected values are displayed

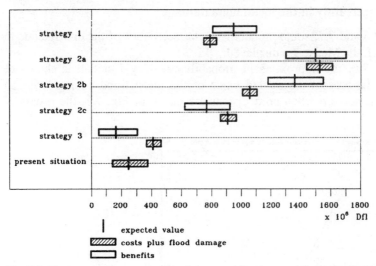

**Fig. 7.8** The present discounted value of the costs plus the remaining mean flood damage and of the benefits over an unbounded horizon (in millions of Dutch guilders). For the present situation and for five strategies, the 80 per cent probability bars and the expected values are displayed

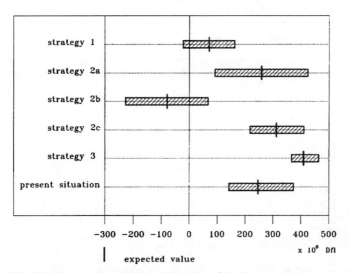

**Fig. 7.9** The net present discounted value of the loss: the costs minus the yields plus the remaining mean flood damage over an unbounded horizon (in millions of Dutch guilders). For the present situation and for five strategies, the 80 per cent probability bars and the expected values are displayed

the uncertainty analysis are displayed in the form of 80 per cent probability bars (i.e. probability masses between the 10th percentile and the 90th percentile). For the present situation and for the five strategies, these bars represent the uncertainty in $s(w, d)$: the present discounted value of the remaining mean flood damage over an unbounded horizon. As should be expected, all strategies result in a smaller expected mean flood damage, and a smaller uncertainty, than for the present situation.

### 7.4.2 Uncertainty in the loss

In Figs 7.8 and 7.9, the uncertainties are displayed in the costs of decisions plus the remaining mean flood damage, the benefits of decisions and the loss. The costs of decisions, i.e. $c(w, d)$, consist of the present discounted value of the costs of extracting sand and gravel, the costs of constructing embankments and dykes, the future cost of cleaning up the polluted summer bed in the north of Limburg, and the remaining costs, over a bounded horizon of 15 years (see Sections 7.3.4 and 7.3.5). The benefits of decisions, i.e. $y(w, d) + s(w, \varnothing) - s(w, d)$, consist of the present discounted value of the yields of extracting sand and gravel over a bounded horizon of 15 years and the reduction in the mean flood damage over an unbounded horizon when choosing decision $d$.

## 7.5 Conclusions

In this paper, we have compared five strategies (combinations of decisions) that reduce damage due to flooding of the Dutch river Meuse. Because we seek an optimum balance between initial investment costs and future flood damage costs, we have used the cost based criterion of the (discounted) loss, where the loss is defined as the net present discounted value of the costs of decisions minus the yields of decisions plus the mean flood damage over an unbounded time-horizon. The strategies have been compared with respect to three types of decision criteria: the criterion of minimal expected loss, the criterion of minimal uncertainty in the loss, and the criterion of maximal safety. On the basis of the results in Section 7.4, we can conclude the following:

- In the present situation, there is a large uncertainty in the present discounted value of the mean flood damage over an unbounded horizon.

- The uncertainties in the losses are relatively large for all strategies.

- The expected present discounted value of the mean flood damage over an unbounded horizon is 'minimal' for the strategies 1, 2a and 2b. These strategies are 'optimal' with respect to the criterion of maximal safety and have a small uncertainty in the flood damage.

- The classification of the strategies on the basis of the expected loss, from a small loss to a large loss, is: strategy 2b – strategy 1 – present situation – strategy 2a – strategy 2c – strategy 3.

- The classification of the strategies on the basis of the uncertainty in the loss, from a small uncertainty to a large uncertainty, is: strategy 3 – strategy 1 – strategy 2c – present situation – strategy 2b – strategy 2a.

- The strategy with minimal uncertainty in the loss is the most expensive strategy: strategy 3 (the construction of embankments and dykes only).

- The extraction of sand and gravel is very uncertain.

- From the viewpoint of both the expected loss and the uncertainty in the loss, even the present situation is to be preferred to strategy 2a.

- The uncertainties in the benefits are larger than the uncertainties in the costs of decisions plus the remaining mean flood damage.

- In comparison with the present situation, the strategy with minimal expected loss (strategy 2b) has a large uncertainty in the loss. For strategy 2b, the uncertainty, and therefore the risks, will be 'shifted' from private persons and industries dealing with flood damage to the government.

# References

Barlow, R.E. 1995: A Bayesian approach to the analysis of reliability data bases. In Balakrishnan, N. (ed.), *Recent advances in life-testing and reliability*. Boca Raton, FL: CRC Press.

Barlow, R.E. and Pereira, C. 1993: Influence diagrams and decision modelling. In Barlow, R.E., Clarotti, C.A. and Spizzichino, F. (eds) *Reliability and decision making*. London: Chapman & Hall, 87–99.

Berger, J.O. 1985: *Statistical decision theory and Bayesian analysis*, 2nd edn. New York: Springer-Verlag.

Bernardo, J.M. and Smith, A.F.M. 1994: *Bayesian theory*. Chichester: John Wiley.

Cooke, R.M. 1991: *Experts in uncertainty; opinion and subjective probability in science*. Oxford: Oxford University Press.

Cooke, R.M. 1994: Parameter fitting for uncertain models: modelling uncertainty in small models. *Reliability Engineering and System Safety* **44**, 89–102.

Cooke, R.M. 1995: *UNICORN: Methods and code for uncertainty analysis*. Delft University of Technology.

Cooke, R.M. and Waij, R. 1986: Monte Carlo sampling for generalized knowledge dependence with application to human reliability. *Risk Analysis* **6**(3), 335–43.

de Finetti, B. 1937: La prévision: ses lois logiques, ses sources subjectives. *Annales de l'Institut Henri Poincaré* **7**, 1–68.

DeGroot, M.H. 1970: *Optimal statistical decisions*. New York: McGraw-Hill.

de Jonge, J.J., Kok, M. and Hogeweg, M. 1996: Modelling floods and damage assessment using GIS. In K. Kovar and H.P. Nachtnebel (eds) *Application of geographic information systems in hydrology and water resources management*. Proceedings of the Vienna Hydro-GIS Conference 1996. Oxfordshire: International Assosication of Hydrological Sciences (IAHS), 299–306.

Delft Hydraulics 1994a: *Onderzoek Watersnood Maas*. Hoofdrapport: De Maas meester (Investigation of the Meuse Flood. Main report: to master the Meuse).

Delft Hydraulics 1994b: *Onderzoek Watersnood Maas*. Deelrapport 2: Beleving van wateroverlast (Investigation of the Meuse Flood. Subreport 2: Experience of flooding).

Delft Hydraulics 1994c: *Onderzoek Watersnood Maas*. Deelrapport 4: Hydrologis-che aspecten (Investigation of the Meuse Flood. Subreport 4: Hydrological aspects).

Delft Hydraulics 1994d: *Onderzoek Watersnood Maas*. Deelrapport 5: Hydraulische aspecten (Investigation of the Meuse Flood. Subreport 5: Hydraulic aspects).

Delft Hydraulics 1994e: *Onderzoek Watersnood Maas*. Deelrapport 9: Schademodellering (Investigation of the Meuse Flood. Subreport 9: Damage assessment modelling).

Delft Hydraulics 1994f: *Onderzoek Watersnood Maas*. Deelrapport 10: Kades, dijken an bijzondere constructies (Investigation of the Meuse Flood. Subreport 10: Embankments, dykes, and particular structures).

Delft Hydraulics 1994g: *Onderzoek Watersnood Maas*. Deelrapport 11: Ontgrondingen (Investigation of the Meuse Flood. Subreport 11: Extracting sand and gravel).

Delft Hydraulics 1994h: *Onderzoek Watersnood Maas*. Deelrapport 13: Ontwikkeling van strategieën (Investigation of the Meuse Flood. Subreport 13: Development of strategies).

Delft Hydraulics & Delft University of Technology 1994: *Onderzoek Watersnood Maas*. Deelrapport 14: Onzekerheidsanalyse (Investigation of the Meuse Flood. Subreport 14: Uncertainty analysis) (DHDUT).

Gottschalk, M.K.E. 1975: *Stormvloeden en rivieroverstromingen in Nederland; deel II, de periode 1400–1600 (Storm surges and river floods in The Netherlands; Volume II, the period 1400–1600)*. Assen: Van Gorcum.

Gottschalk, M.K.E. 1977: *Stormvloeden en rivieroverstromingen in Nederland; deel III, de periode 1600–1700 (Storm surges and river floods in The Netherlands; Volume III, the period 1600–1700)*. Assen: Van Gorcum.

Jae, M. and Apostolakis, G.E. 1992. The use of influence diagrams for evaluating severe accident management strategies. *Nuclear Technology* **99**, 142–57.

Kok, M. 1995: Decision support for assessing flood damage reduction strategies of the Meuse river in the Netherlands. Presented at the *Workshop on Methods for Structuring and Supporting Decision Processes*, Laxenburg, Austria: IIASA.

Mendel, M.B. 1989: Development of Bayesian parametric theory with applications to control. *PhD thesis*, Massachusetts Institute of Technology, USA.

Misiewicz, J.K. and Cooke, R.M. 1997. Isotropy and stochastic rescaling. *Probability and Mathematical Statistics*. (to appear).

Savage, L.J. 1972: *The foundations of statistics*, 2nd edn. New York: Dover Publications.

Shaw, E.M. 1988: *Hydrology in practice*, 2nd edn. London: Chapman & Hall.

van Noortwijk, J.M., Cooke, R.M. and Misiewicz, J.K. 1994: A characterisation of generalised gamma processes in terms of sufficiency and isotropy. *Technical Report 94-65*. Faculty of Mathematics and Computer Science, Delft University of Technology, The Netherlands.

van Noortwijk, J.M. and Peerbolte, E.B. 1995: Optimal sand nourishment decisions. *Publication 493*. Delft Hydraulics, The Netherlands.

## Appendix

- **Definition 1 (gamma distribution).** A random quantity $X$ has a gamma distribution with shape parameter $a > 0$ and scale parameter $b > 0$ if its probability density function is given by:

$$\text{Ga}(x \mid a, b) = [b^a / \Gamma(a)] x^{a-1} \exp(-bx) I_{(0,\infty)}(x)$$

- **Definition 2 (inverted gamma distribution).** A random quantity $Y$ has an inverted gamma distribution with shape parameter $a > 0$ and scale parameter $b > 0$ if $X = Y^{-1} \sim \text{Ga}(a, b)$. Hence, the probability density function of $Y$ is:

$$\text{Ig}(y \mid a, b) = [b^a / \Gamma(a)] y^{-(a+1)} \exp(-b/y) I_{(0,\infty)}(y).$$

- **Definition 3 (beta distribution).** A random quantity $X$ has a beta distribution with parameters $a$, $b > 0$ if its probability density function is given by:

$$\text{Be}(x \mid a, b) = \frac{\Gamma(a+b)}{\Gamma(a)\Gamma(b)} x^{a-1} (1-x)^{b-1} I_{[0,1]}(x)$$

- **Definition 4 (gamma-gamma distribution).** A random quantity $X$ has a gamma-gamma distribution with parameters $a$, $b > 0$ and $n = 1, 2, \ldots$ if its probability density function is given by:

$$\text{Gg}(x \mid a, b, n) = \int_0^\infty \text{Ga}(x \mid n, \lambda) \text{Ga}(\lambda \mid a, b) \, d\lambda$$

$$= \frac{\Gamma(a+n)}{\Gamma(a)\Gamma(n)} \left[ \frac{x}{b+x} \right]^{n-1} \left[ 1 - \frac{x}{b+x} \right]^{a-1} \frac{b}{(b+x)^2} I_{(0,\infty)}(x)$$

# 8 The ABLE story: Bayesian asset management in the water industry [1]

A O'Hagan

## 8.1 Asset management plans

This chapter describes a collaboration with UK water companies in which I have been engaged since 1988. It began when the 10 publicly owned water authorities in England and Wales were to be privatized. The privatization took effect in November 1989, but of course the preparations for this exercise were extensive. In particular, there was considerable public anxiety over the privatization of water and the British government wished to provide some assurance that the newly privatized water companies would manage their businesses responsibly in the public interest. A regulatory body was established, known as the Office of Water Services, or OFWAT. One of the requirements placed on the water authorities in their preparations for privatization was to submit an asset management plan (AMP). The purpose of the AMP was to set out the extent of the assets that the water company would own, their condition and any current problems with their performance; to identify the work that would be required to remedy deficiencies in performance and service, and to maintain the condition and serviceability of assets; to forecast growth in demand for water, and the consequent asset development projects needed to accommodate that growth; to set out the

[1] Many people helped and sustained me throughout this long collaboration. In the water companies, their certifiers and agents I have made many friends, and with apologies to those I leave out I will mention just a few – Edward Glennie, Arthur Boon and Bill Kingdom (WRc), Roger Hocking (South West Water), David de Hoxar (Southern Water), David Stratford (Thames Water), Frank Wells (Anglian Water), Steve Bottom, John Brindley and Warwick Eden (W S Atkins). Several fellow statisticians have worked with me and contributed their advice and expertise – Michael Goldstein, Peter Freeman, Wai Fong Lai, Cliff Litton, Geoff Freeman, Saul Jacka and Wilfrid Kendall. Tim Maskell and Jenny Rowley of Nottingham University Consultants Ltd deserve a mention, and of course my wife and family who have put up with the very many long unsocial hours I have laboured on this work whilst also carrying a full-time university job. My sincere thanks go to all of those.

costs of all this asset maintenance, renewal, improvement and extension; and finally to bring all this together into a fully costed programme of asset management over a 20 year period.

The AMP was a massive and ambitious undertaking for each company and, as stated above, was quite unachievable because no company could know all the detailed information it called for, and nobody could forecast perfectly what will happen over 20 years. It was recognized at the outset by OFWAT that statistical expertise would be needed and that very many aspects of the AMP could only be estimated. OFWAT's guidelines for the preparation of the AMP included requirements to describe the statistical uncertainty in all estimates.

A water company's assets are very extensive, and a large proportion are buried underground in the form of literally millions of km of water pipe and sewers and associated structures. The company could not know definitely the condition of most of these pipes without physically digging them up. The expense and disruption which would be caused by digging up even a tiny fraction of a company's pipes would be phenomenal and even less intrusive investigations would be quite impractical to apply throughout a company. Public concern focused partly on the difficulty of ensuring that the new water companies would not simply allow underground assets to decay, unnoticed and unattended, and so the AMP was initially introduced as a requirement to plan for underground assets only. However, it was soon extended to include above-ground assets such as water treatment works, service reservoirs (e.g. water towers), pumping stations and sewage treatment works.

Some other basic facts about the AMP and the privatization exercise will illustrate its importance to the water authority (and subsequently to the private water company). A primary function of OFWAT is to place limits on the charges that water companies can make to their customers. Specifically, OFWAT set, for each of the 10 companies at privatization, a factor known as '$K$', which specified the maximum rate of increase the company could apply each year to its water charges. Thus, a company given a '$K$' of 10 per cent would be allowed to raise its charges by 10 per cent every year. (This is in addition to any agreed general inflationary increase, so it means a 10 per cent increase in real terms.) In principle, a negative $K$ would force a company to reduce its prices annually, but it was recognized that the water industry had been chronically underfunded for years and so some increases were inevitable. As publicly owned utilities the water authorities' charges were subject to political pressures. Indeed it is not completely unfair to suggest that a motivation for the government in privatizing the industry and in requiring it to estimate fully its investment needs for the future, was to make it politically possible to increase prices enough to remedy past underfunding.

In effect, the AMP is the company's bid to OFWAT to be awarded a high $K$. It is therefore important for OFWAT to be assured of the quality and honesty of a company's estimates. Each company was required to appoint an independent firm of engineering consultants with expertise in the water industry as their certifiers. All the data used by the company, and all the methodologies used to obtain those data, had to be approved and audited by their certifiers. Furthermore the AMP was to form part of each new company's prospectus for their stock market flotation. If investors in 'the City' were to have the impression that a company's plans were in any way suspect, or based on unreliable or untried methodology, it might seriously affect the sale of their shares.

Following privatization in 1989 each of the 10 water companies was required by OFWAT in 1994 to submit a second AMP, known informally as AMP2, as part of a revision of $K$. At the time of writing this chapter (1995), the companies are anticipating being required to do AMP3, perhaps in 1999 after another 5 years. However, they are still adjusting to the effects of a general reduction in $K$ at this round, and have not begun seriously to think about having to do it again!

In the run-up to privatization, I worked for two water authorities, South West Water and Anglian Water. In preparing for AMP2, I worked for these two companies again, together with Southern Water, Thames Water and North West Water. I also worked for two companies which were already private companies in 1989, Bristol Water and Three Valleys Water. In contrast with the 10 large, newly privatized companies these are two of a number of 'water only' companies, who are responsible for the provision of clean water over a (usually relatively small) area but not for the removal and treatment of sewage. Finally, I also assisted the Water Service of the Department of the Environment, Northern Ireland. Being outside England and Wales, this undertaking was still publicly owned but had decided that it would also benefit from preparing an AMP, and this happened to be at the same time as the English and Welsh companies were developing their AMP2s.

This chapter will describe how, through this long association with clients who face a large and complex problem, in which uncertainty plays a key role, I have developed a rather specialized Bayesian methodology. The details of that methodology have been published in a series of papers, and those will be referred to as appropriate. However, my intention here is to concentrate on the interaction with the clients. I will describe how the methodology evolved, always in response to the clients' evolving needs, and to my growing understanding of their industry and the statistical implications of these needs. I will also deal with the problems of implementing this methodology, involving finding effective ways for the company's personnel to identify specific models and to specify their prior information.

The chapter is organized chronologically, so that the evolution of the methodology and its application (and the interaction with the client at each stage) can be clearly seen. I hope that this will thereby represent a useful case study in the application of the Bayesian paradigm.

## 8.2  Why a Bayesian approach?

This section deals with why I originally proposed a Bayesian approach, and how it came to be accepted in 1988 by South West Water and Anglian

Water. The cynical answer to the first part is that I *always* propose Bayesian methods because I am so firmly committed to the Bayesian paradigm! It is true that I am a firm proponent of Bayesian methods and that I tend to see scope for Bayesian analysis in every problem: that is how I am. But it does not serve the cause of Bayesian statistics well to persuade someone to use a Bayesian approach when its use is not justified. It takes more effort to do a Bayesian analysis. If there is strong prior information or a complex data structure that only Bayesian computational methods can tackle, the rewards can easily outweigh the cost and one can then recommend a Bayesian approach with due enthusiasm. If, on the other hand, there are plenty of data and relatively weak prior information, and a straightforward data structure leading to a classical analysis that is known to be sound (i.e. approximating to a proper Bayesian analysis in the case of weak prior information), then there is no point in beating the Bayesian drum too exuberantly.

I was first approached in early 1988 by Severn-Trent Water. At that time I was at the University of Warwick, and we had created a Statistical Consultancy Unit in order to seek interesting practical problems for academic statisticians to work on (and to make a bit of money!). I believe that challenging practical problems provide a powerful impetus for theoretical development, and this case study serves partly to illustrate how fruitful the interaction can be. It is also important to remember, however, that the clients have different priorities. They may be interested in principle in theoretical developments, because they may make their life easier in the long term. But they also face the immediate problem which requires some kind of solution in the short term (and often the time scale is *very* short!). The consultant must primarily address the problem at hand in an efficient and business-like way – qualities which should apply to all the statistician's dealings with the client throughout the project. The Statistical Consultancy Unit at Warwick University was created with these considerations in mind, to present a professional service to industry.

I and my colleague, Dr Geoff Freeman, who was director of the consultancy unit, visited Severn-Trent to find out about the problem, since at that time neither of us knew anything about the water industry or the requirement for an AMP that was then emerging in the industry. The problem as it was presented to us was one of classical survey sampling inference. Remembering that the expectation at that early stage was for an AMP based on underground assets alone, the authority proposed to divide their region into reasonably self-contained zones. A zone for water distribution might contain the network of water mains taking water from a single service reservoir (where the clean, treated water is stored) to nearby customers. A zone for sewage disposal might consist of the network of sewers (known as 'the sewerage') draining and collecting waste water to be treated by a single sewage treatment works. It was proposed that we should randomly sample zones, that the authority would estimate costs for maintaining, replacing or improving the underground systems in the sampled zones and that we should thereby estimate the total cost. The sampling scheme might include stratification or other standard devices of survey sampling but the final estimate would be a standard classical estimator of the population total.

It became clear to me, however, that (1) each zone study, yielding a single observation, could be *very* costly, and (2) the authority had substantial prior information. They might not have known the detailed condition of their pipes – indeed, they did not even know where they all were – but they did have a good idea of how they performed generally. On the basis of prior information they could, for instance, confidently expect to spend more on replacing water mains to reduce the incidence of burst mains in certain zones than in others.

Geoff Freeman and I presented a proposal in competition with two other universities, but did not get the job. (It was awarded to Sheffield University.) Our joint proposal was that Geoff would deal with the water distribution side, using classical methods, while I would apply a Bayesian approach for sewerage. (We understood that sewerage studies were more expensive, but that relatively good prior information could be found.) Although the proposal was not accepted, one of the people from Severn-Trent who had been at our presentation of our proposals wrote privately to me afterwards. He said that he thought the Bayesian approach was very interesting, and that I might try to see whether another water authority would wish to use it. He suggested that I might contact Dr Edward Glennie at WRc (formerly the Water Research Centre).

WRc are the water industry's own research organization, whose work was then funded partly by subscriptions from all the water authorities and partly by contracts for specific projects with individual authorities. Edward Glennie is one of their more experienced statisticians and I discovered that he had been influential in advising OFWAT to recognize the importance of good statistical input to the AMP exercise. It was on the advice of WRc that Severn-Trent (and other water authorities) had decided to use a classical sample survey methodology. Edward agreed to see me and was sufficiently impressed with my ideas that he decided to recommend to South West Water that they should consider using a Bayesian approach. WRc had a contract as consultants to South West Water to prepare their AMP.

In view of the importance to them of their AMP, and of the risks in being seen to use inferior methods in its preparation, it is not surprising that South West Water were cautious. A presentation I made to some of their senior planners was followed by another to their certifiers, W S Atkins Ltd. The certifiers sought advice from another group of professional statisticians in industry, who posed a number of perceptive questions. Eventually all sides were agreed and the Statistical Consultancy Unit were engaged to provide a statistical analysis in support of South West Water's AMP. It was pleasing (and somewhat daunting!) to me to have created an opening for Bayesian methods in such a high-profile context, where the sums of money which were expected to emerge as the final estimates would run to billions of pounds.

Both South West Water and W S Atkins soon developed quite a strong commitment to the Bayesian approach. Just a couple of months after I had started to work for South West Water, I was invited to develop the same methods for Anglian Water. They had been strongly advised to follow this approach by W S Atkins who were also Anglian's certifiers.

## 8.3    An approach based on Bayes linear methods

So what was the Bayesian method that I had been selling to WRc, South West Water and W S Atkins? It was certainly not a fully fledged technique at that stage, but was based on some research I had recently done. When asked to teach an undergraduate course on sample surveys, I had wanted to include some reference to Bayesian methods alongside the standard classical theory. Here, even more than other areas of statistics, the Bayesian paradigm seemed to require the user to enter into much more detail and complexity than classical theory. The point was that the classical methods only concerned themselves with best unbiased linear estimators, and were based on a finite population analogue of the Gauss–Markov theorem. This was all non-parametric, in the sense that no distributions were assumed. Fully Bayesian methods would need to specify a high-dimensional prior distribution. However, I knew of the existence of a Bayesian analogue of Gauss–Markov theory – Bayes linear methods. There had been a little Bayesian theory in sample surveys developed along these lines (see, for instance, Ericson, 1969; Smouse, 1984), but I took this a bit further and assembled it into a more comprehensive form in a paper (O'Hagan, 1984). Another paper (O'Hagan, 1987) applied the ideas to randomized response.

The application of Bayes linear methods in sample surveys is straightforward, and proceeds as follows. Consider a finite population of $N$ individuals and let $x_i$ be the unknown characteristic of interest for individual $i$. Then one wishes to make inference about the vector $\mathbf{x} = (x_1, x_2, \ldots, x_N)'$ or about some function of $\mathbf{x}$. The population total, for instance, is $t = \sum x_i = \mathbf{1}_N' \mathbf{x}$, where $\mathbf{1}_N$ is a vector of $N$ ones. Let the vector of prior expectations of the $x_i$s be $\mathbf{m}$ and the matrix of prior variances and covariances be $\mathbf{V}$. Now suppose that one observes a sample so that one learns the value of the characteristic of interest for each of the individuals in the sample. Suppose for convenience that the sample members are the first $n$ individuals in the population so that the observation vector is $\mathbf{y} = (x_1, x_2, \ldots, x_n)'$. Denote by $\mathbf{z} = (x_{n+1}, x_{n+2}, \ldots, x_N)'$ the remaining $x_i$s which are still unknown. So we have divided $\mathbf{x}$ by $\mathbf{x}' = (\mathbf{y}', \mathbf{z}')$ and we now divide $\mathbf{m}$ and $\mathbf{V}$ accordingly:

$$\mathbf{x} = \begin{pmatrix} \mathbf{y} \\ \mathbf{z} \end{pmatrix}, \qquad \mathbf{m} = \begin{pmatrix} \mathbf{m}_y \\ \mathbf{m}_z \end{pmatrix}, \qquad \mathbf{V} = \begin{pmatrix} \mathbf{V}_y & \mathbf{W} \\ \mathbf{W}' & \mathbf{V}_z \end{pmatrix}$$

Then the Bayes linear estimate (BLE) of $\mathbf{z}$ is

$$\hat{\mathbf{z}} = \mathbf{m}_z + \mathbf{W}' \mathbf{V}_y^{-1} (\mathbf{y} - \mathbf{m}_y)$$

which in Bayes linear theory is thought of as analogous to the posterior mean of $\mathbf{z}$ in a full Bayesian analysis. The analogue of the posterior variance-covariance matrix is the Bayes linear dispersion matrix

$$\mathbf{D}_z = \mathbf{V}_z - \mathbf{W}'\mathbf{V}_y^{-1}\mathbf{W}$$

If we now wish to estimate the population total $t$, its BLE is

$$\hat{t} = \mathbf{1}_n' \, \mathbf{y} + \mathbf{1}_{N-n}' \, \hat{\mathbf{z}}$$

with dispersion

$$D_t = \mathbf{1}_{N-n}' \mathbf{D}_z \mathbf{1}_{N-n}$$

The Bayes linear approach can reproduce all the basic classical sample survey theory, including the cases of stratified sampling or cluster sampling, as (limiting forms of) special cases. However, the Bayesian approach is very much more flexible because through the prior mean vector $\mathbf{m}$ and variance-covariance matrix $\mathbf{V}$ one can represent genuine, substantial prior information. There is another important difference, which is that, in classical theory, what determines the estimator one uses is the random sampling method, whereas in Bayesian inference it is the prior distribution (specified in limited form as $\mathbf{m}$ and $\mathbf{V}$) which determines the inferences. For instance, if one uses simple random sampling then the standard classical estimator of $t$ is $N\bar{y}$, where $\bar{y}$ is the sample mean. If stratified sampling were used then a different estimator should be used. This is odd, because suppose for instance that one investigator took a simple random sample and another investigator took a stratified sample from the same population, and by chance chose exactly the same $n$ individuals as the first investigator. Both would obtain exactly the same data but they would make different inferences.

In the Bayesian approach it does not matter how the sample was chosen (provided that the selection mechanism is formally 'non-informative' – see for instance Sugden and Smith, 1984), only what data are observed. It is not even strictly necessary to choose the sample randomly at all. In fact the optimal approach for a Bayesian wishing to estimate $t$ should be to choose the sample to minimize $D_t$. (Since $D_t$ does not depend on $\mathbf{y}$, the sample can indeed be chosen in this way.)

To apply the basic theory to the AMP situation, the population is the set of $N$ zones and $x_i$ is the investment cost needed over the 20 year AMP period for refurbishment, improvement and extension of assets in zone $i$. The objective is to estimate the total investment cost over all zones, i.e. $t$. The theory then requires us to supply prior means, variances and covariances for the $x_i$s, to make up the vector $\mathbf{m}$ and the matrix $\mathbf{V}$. At the time when I was trying to persuade South West Water and their certifiers that a Bayesian approach would be viable, I had no firm ideas of how I would elicit this prior information from the water authority personnel (and of course that is a difficult task), but I knew that the basic theory was much more powerful and flexible than the classical methodology which was their alternative.

## 8.4    Modelling a zone-based prior

Having satisfied both South West Water and W S Atkins that my basic approach was both viable and potentially very powerful, I had to put it into practice. As already remarked, this meant finding a way to elicit prior information about the zone costs $x_1$, $x_2$, ..., $x_n$. The answer to this problem lies in the principle of *elaboration* which I developed in O'Hagan (1988), and in this context consists of observing how the water authority personnel would construct a prior estimate of $x_i$ if required to do so. The following emerged from a series of discussion meetings between myself and the WRc staff who formed the team of consultant engineers that was developing South West Water's AMP.

Consider the estimation of water distribution costs so that $x_i$ was the investment cost in one of South West Water's $N = 59$ water distribution zones. When I asked the engineers how they would estimate $x_i$ they said they would begin by thinking about the various kinds of work that might need to be done to the system. They identified five principal areas of expenditure:

1. repair or replacement of mains that are found to be in poor structural condition;

2. expansion of the mains network to accommodate increased demand for water within the zone;

3. scraping and relining of mains found to be in poor internal condition;

4. expansion of capacity of service reservoirs to meet increased demand for water;

5. connecting newly built properties into the network over the AMP period.

Several other areas of expenditure were discussed, but were expected to have much lower cost implications than these five.

I then asked how the engineers would estimate each of these costs in turn. In the case of the first area of expenditure, for instance, they said it would depend on how big the zone was, in terms of the total length of mains, what the overall structural condition of mains in that zone was, and the general cost per metre of refurbishing (i.e. repairing or replacing as necessary) the poor mains. They could readily estimate the total length of mains in a zone, and indeed could simply ask South West Water to tell them what the length of mains in a zone was according to their records. It was acknowledged that the records were far from accurate, but they would at least provide an estimate. The other factors were more difficult to access. The overall structural quality of mains in a zone could not be quantified, but on the basis of records of burst mains and the local knowledge of South West Water engineers it was felt possible to divide zones crudely into three structural condition strata – 'good', 'average' and 'poor' condition. There was reasonable historic information on the unit cost, in pounds per metre, of laying new pipes of given diameters or of repairing the larger pipes, and depending on whether the pipes lay under a road (which would thereby entail a cost of digging up the road), under a grass verge or a field. But it was not known what proportions of poor mains would be found to be of any given diameter or lying under road, verge or field. It was eventually decided to express the cost for mains refurbishment in zone $i$ as a product $M_i(SM)_i$, where $M_i$ is the total length of mains in zone $i$ and $(SM)_i$ is a unit cost per metre of *total* mains (*not* per metre of *poor* mains) for refurbishment. Then $(SM)_i$ would be estimated at £1/m in a 'good' zone, £2/m in an 'average' zone and £3/m in a 'poor' zone. These are obviously crude estimates but were the best that the engineers could come up with without engaging in detailed investigation of each zone.

The other four areas of expenditure were similarly analysed, and an allowance made for the neglected 'other costs'. Details of the resulting model for estimating $x_i$ are given in O'Hagan *et al.* (1992; hereinafter referred to as O'HGB).

Having structured and developed prior estimates of the $x_i$s in this way, it is natural to consider prior uncertainty about the $x_i$s (as expressed in their prior variance-covariance matrix $\mathbf{V}$) as arising from uncertainty about the components of the model. For instance, uncertainty about costs for mains refurbishment arises from uncertainty about $M_i$ and $(SM)_i$. $M_i$ and $(SM)_i$ were two of the 11 quantities which appeared in the full model for $x_i$ given in O'HGB. I now needed to consider eliciting variances and covariances of these 11 quantities in each zone, i.e. $11 \times 59 = 649$ quantities altogether. That implies the enormous task of eliciting 649 variances and 210 276 covariances. Some further simplification was essential!

The first step is to use independence. I have always thought that to a Bayesian who adopts a subjective view of probability, the concept of independence is particularly simple, and I always explain it in this way: 'If when you learn something about one quantity your beliefs and expectations about likely values of another quantity are unchanged, then those quantities are independent'. With independence clearly understood in this way, the engineers were able to assert that $M_i$ and $(SM)_i$ were independent: if they learnt that a zone had more mains than their prior expectation, then it would not alter their beliefs about the unit cost for refurbishment of mains in that zone, or in any other zone. They were in fact willing to accept independence between all the 11 quantities in the basic model (and, to some extent the modelling of their beliefs about zone costs was guided by my recognition that independence would be an important simplifying feature).

The elicitation problem is now reduced to needing to obtain a $59 \times 59$ variance-covariance matrix for each of the 11 quantities separately. This is still a major task, and could not be greatly simplified by further independence assumptions. If we look at the mains lengths $M_1, M_2, \ldots, M_{59}$, for instance, the definition makes it clear that these should not be independent. If the company learns that the length of mains in zone $i$ is more than the prior expectation, then its expectations for the lengths of mains in other zones will increase. If they have underestimated in one zone, this suggests a general tendency to underestimation in their method for obtaining prior estimates. In some other quantities, however, some partial independence was assumed. Thus for the unit costs $(SM)_i$, the engineers were prepared to say that the values for two zones that were in different strata would be independent. (This was at best an approximation, and was relaxed in later work.)

Even the smallest subproblem remaining is still complex. For instance, just 14 zones were judged to be of 'poor' overall condition, so the variance-covariance matrix of the $(SM)_i$ values for these zones has 14 variances and 91 covariances to elicit. Fortunately, further simplification is possible because these 14 values could be regarded as exchangeable. Having used their judgement to classify these 14 zones as 'poor', the engineers cannot distinguish between them in respect of refurbishment unit costs. In particular, they have no reason to believe that the unit cost will be higher in one particular zone than in another, given that both are classified as poor, and this is reflected in giving a prior expectation of £3/m to each. Exchangeability means that they all have the same variance and all pairs have the same covariance. Exchangeability creates a dramatic simplification, so that I now need elicit only one variance and one covariance.

As if never satisfied, I still looked for more simplification, because eliciting a covariance seems like a difficult task to me. We can represent the exchangeable quantities in the following way. If $y_1, y_2, ..., y_{14}$ are the values of $(SM)_i$ in the poor zones, we can write

$$y_i = \mu + e_i$$

where $\mu, e_1, e_2, ..., e_{14}$ are all independent. This represents each $y_i$ as an overall average unit cost $\mu$ plus a disturbance $e_i$ expressing how this zone's unit cost deviates from the average $\mu$. It is uncertainty about the overall average $\mu$ which accounts for covariance between the zone unit costs, for if $\text{Var}(\mu) = v$ and $\text{Var}(e_i) = w$, then $\text{Var}(y_i) = v + w$ and $\text{Cov}(y_i, y_j) = v$. So we can elicit the variance and covariance by separately eliciting the two *variance components* $v$ and $w$. This model is particularly appropriate because it captures the essential reason why the unit costs in different zones should be correlated in the first place. If South West Water learnt, for instance, that the unit cost in one of the poor zones was, say £2.50/m instead of the prior expectation of £3, it suggests that they should now expect lower unit costs in other poor zones precisely *because* they would suspect that £3 was too high as an estimate of the overall average $\mu$. Furthermore, the extent to which beliefs about other poor zones are affected by the observation of £2.50 depends on the relative magnitudes of $v$ and $w$. For if $v$ is much larger than $w$, reflecting great uncertainty about the overall level but a belief that unit costs vary relatively little from zone to zone, then one would now expect to see unit costs close to £2.50/m in other zones. At the other extreme, $w$ much larger than $v$ corresponds to an expectation of wide variations of unit costs between zones and a belief that £3/m is relatively accurate overall, in which case unit cost estimates for other zones would change very little from the prior estimate of £3/m.

Before attempting to elicit the variance components, I discussed the above interpretation with the engineers, with the view of helping them to understand the significance of the numbers I wanted from them. I also discussed variance components in terms of signal (variation in $\mu$) versus noise (variation of $e_i$s). The better the signal-to-noise ratio, the more strongly we learn from new data. Fortunately, I did not also have to explain the concept of a variance or standard deviation, since some of the WRc personnel had enough statistical training to have a good intuitive appreciation of standard deviation. For the $(SM)_i$ values in poor zones, I first asked how close to the estimated £3/m the real average unit cost might be, averaged over a large number of poor zones. A standard deviation of 0.25 was elicited ($v = 0.25^2$). I then asked how close to the estimated £3/m the real unit cost might be in an individual poor zone *assuming* that £3/m was correct on average. This amounted to asking about how much the unit costs varied between zones. (I did not choose the latter wording for two reasons: first because it invites confusion with the standard deviation of the difference between two zones, which is $\sqrt{2}w$, and second because the former wording is more appropriate in other cases, as we shall see with the mains lengths.) A standard deviation of 0.25 was elicited for this also ($w = 0.25^2$). The same procedure was used for eliciting the variance-covariance matrices for $(SM)_i$ values in the 'average' and 'good' strata.

Turning to the mains length values $M_i$, we cannot use the same approach exactly because the $M_i$ are clearly not exchangeable. The prior expectation of mains lengths varied greatly from about 4 km in the smallest zone to 545 km in the largest. Furthermore, the variances could not be equal, since the variance in the smallest zone must certainly be less than in the largest. For such quantities, I found that the engineers naturally expressed their uncertainty in terms of percentage errors. This suggested that if we considered instead the random variables $M_i^* = M_i/E(M_i)$, where each length was divided by its prior expectation, then the $M_i^*$ would become exchangeable (with unit expectations). If they had variances $v + w$ and covariances $v$, then this results in $\text{Var}(M_i) = (v + w) \, E(M_i)^2$ and $\text{Cov}(M_i, M_j) = vE(M_i) E(M_j)$. So I proceeded to elicit variances in terms of an overall average percentage error $v$ and a percentage error $w$ for each zone (assuming the overall average would be correct). The exchangeability model proved to be adequate to represent the engineers' beliefs about all the quantities in the model, with the use of stratification of zones and/or scaling by dividing by the prior estimate as appropriate for each quantity. Details are given in O'HGB.

The same approach was used for modelling prior beliefs about costs in sewerage zones, and also for Anglian Water, but it should be stressed that in each case a separate model for zone costs needed to be constructed, and then variance components elicited for each quantity in the model. The model used for sewerage zones in Anglian Water is described in O'Hagan and Wells

(1993, hereafter referred to as O'HW). This paper also provides much more explanation of the background to the AMP and how the model was built than O'HGB.

## 8.5   The nature of data

Having solved the problem of building the prior, I felt ready to receive the sample data and to apply the Bayes linear method to derive posterior inferences. I already knew that the Bayes linear estimator of Section 8.3 was not strictly applicable because the data do not consist of perfect observations of the costs in the sample zones. The data are the results of intensive engineering study of individual zones but are still merely good estimates of zone costs. So it was necessary to allow for errors in the data, but I thought that would be straightforward. If we let the observation vector $z$ equal $(x_1, x_2, \ldots, x_n)$ *plus* some independent error, it is quite simple to derive a new Bayes linear estimate, for the whole of $x$ since now we do not know any $x_i$ exactly. For the studied zones, for instance, the posterior estimate will be a weighted average of the study estimate $z_i$ and the prior expectation $m_i$, of a form familiar in Bayesian statistics.

I worked through some simple numerical examples and presented them to South West Water, but received an unexpected response. They wanted the posterior estimate to be the study estimate $z_i$. They argued that if they had just spent many thousands of pounds to produce this estimate, they were not going to partially throw it away and allow it to be influenced by the much less well considered prior estimate. I thought initially that their reaction was irrational but was soon convinced that they were justified. The point is that a zone study *incorporates* all the prior knowledge, as well as all the additional information revealed in the study itself, and so is really a posterior estimate.

I therefore had to adjust my basic theory somewhat. The solution is described in O'HGB and in O'HW. As the latter explains, I was not entirely satisfied with the solution, and this matter is taken up again in Section 8.7 below. A similar procedure to that used to model prior information was employed to quantify the error variances of the zone studies, and this is also described in O'HGB and O'HW.

## 8.6   Transition

The methods developed so far were successfully applied in the preparation of AMPs for South West Water and Anglian Water, and the 10 companies were duly privatized in 1989. The $K$ values allowed them by OFWAT were unsurprisingly high and the new water companies embarked upon massive investment programmes to remedy the historic underfunding and to improve the service to customers. The companies were in transition to a new, more commercially oriented existence. There followed a period of transition for me too, beginning with a move in 1990 from Warwick to Nottingham University and the acquisition of the title of 'Professor'.

As far as I was concerned, the water companies went quiet for about 2 years, for which I was grateful in view of my new responsibilities. However, in 1991 the ball was set rolling again by another call from Severn-Trent Water. OFWAT had given notice that AMP2 was to be required, and it was likely to be an even bigger exercise than before. As it happens, Severn-Trent again went elsewhere. (They definitely wanted to use the Bayesian approach, but chose to engage Warwick University's Statistical Consultancy Unit, where Geoff Freeman, Dr (later Professor) Jim Smith and Dr Saul Jacka had helped me at privatization.) But over the next year or so I acquired several new clients. It was time to review the methodology, both to remedy perceived defects in the original approach and to meet the new imperatives of AMP2.

One difficulty had arisen even before privatization, and concerned the fact that the methodology I had created was only capable of estimating or providing inference about total costs for zones, and for functions of these like the grand total cost. This was of course all that was originally asked for, but it was not long before a need for more detail emerged. One of the many last minute additions to OFWAT's requirements for the AMP was to break down cost estimates by cause. They wanted to know what proportions of costs were attributable to maintaining asset condition and performance at current levels, improving the assets and their performance to meet higher levels of service and accommodating increased demand. Although I produced a rather *ad hoc* procedure at the time, I was not happy with it. The need for a proper solution was emphasized soon afterwards, when South West Water wanted to break down zone cost estimates to give estimates for different kinds of work. A company naturally would like such information in order to plan the allocation of resources, but the original methodology was not designed to deliver such detail. The need became imperative when it began to emerge that OFWAT would be demanding much more detailed estimates in the companies' AMP2 submissions.

The solution was in fact a natural development from my earlier approach. I had developed with each company a model for zone costs, and elicited means, variances and covariances for all the quantities in the model in order to obtain means, variances and covariances for zone costs so that I could apply the BLE. Yet now the quantities within the model also needed to be estimated. The solution was not to apply the BLE to zone costs but to the various quantities within the model. I already had their prior means, variances and covariances. The only extra thing required was that when conducting the detailed sample studies of zones, the engineers should return not just a single estimate of the total zone cost but estimates of all the separate quantities in the model.

Although simple in principle there were some important consequences of this shift of the inferential focus from zone total costs to the various components of the cost.

The first consequence was in the construction of the model itself. Previously, the modelling was used to build prior information for zone costs, and as such the quantities within it were not significant in their own right. Quantities were introduced simply to reflect how engineers would think about estimating zone costs. For AMP2 the quantities within the model had to include all those things which the company wanted to estimate. At the outset it was a matter of what the company themselves *wanted*, but with the publication of OFWAT's guidelines for AMP2 the modelling was driven more by what the company *needed* to estimate. Models became more elaborate, and were broken into a series of equations, which would typically be thought of as breaking the grand total cost down into more and more detail. The company wanted to estimate individual quantities appearing at the bottom of this cascade of equations, plus the grand total cost at the top, and usually many of the intermediate quantities too. In fact, equations might be added purely as 'side calculations' to define extra quantities of interest in terms of others already defined in other equations.

As an example of the modelling used in AMP2, the following might be equations in a (rather simplified) model of water distribution costs:

$$\text{Total} = \Sigma_j \text{ DivCost}_j \qquad (8.1)$$

$$\text{DivCost}_j = \Sigma_{i(j)} \text{ ZoneCost}_i \qquad (8.2)$$

$$\text{ZoneCost}_i = \text{RehabC}_i + \text{Pressure}_i + \text{SysGrow}_i + \text{New}_i + \text{Develop}_i \qquad (8.3)$$

$$\text{RehabC}_i = \text{Length}_i \times \text{RehabUC}_i \qquad (8.4)$$

$$\text{Length}_i = \text{Iron5}_i + \text{Iron4}_i + (\text{Iron3}_i \times \text{Decay}_i) \qquad (8.5)$$

$$\text{Pressure}_i = \text{Lowprops}_i \times \text{PresCost}_i \qquad (8.6)$$

$$\text{SysGrow}_i = \text{IncDemand}_i \times \text{GrowCost}_i \qquad (8.7)$$

$$\text{IncDemand}_i = \text{Growth}_i + \text{NewDemd}_i - \text{LeakRed}_i \qquad (8.8)$$

$$\text{New}_i = \text{NewProps}_i \times \text{Connect}_i \qquad (8.9)$$

$$\text{Develop}_i = \text{Newprops}_i \times \text{NewFac}_i \times \text{OnOffC}_i \qquad (8.10)$$

Equation (8.1) defines the total cost as a sum of costs in each division of the company, and eq. (8.2) defines divisional costs as sums of zone costs (where $\Sigma_{i(j)}$ denotes summing over those zones $i$ which are in the division $j$). Equation (8.3) splits zone costs into various components, and the remaining equations further express each component in more detail. For instance, eq. (8.4) defines the rehabilitation cost for poor mains as a length times a unit cost, reminiscent of the example used earlier. This time $\text{Length}_i$ is the actual length of mains needing rehabilitation (rather than the length of all mains in the zone), which (8.5) defines using the concept of condition grading. A pipe in condition grade 1 is (as good as) new, whereas one in grade 5 is seriously corroded and should be replaced as soon as possible. Grades 2 to 4 have appropriate meanings between these extremes. The modelling shows the company taking a proactive role in improving condition, and eq. (8.5) accordingly defines the length needing rehabilitation as equal to the length in condition grade 5 plus the length in condition grade 4 plus a proportion Decay$_i$ of the length in grade 3 (reflecting that grade 3 is acceptable but some grade 3 mains will decay to the unacceptable grade 4 over the 20 year period covered by the AMP). These lengths are for iron mains only, since corrosion is not a problem in non-iron mains. (Note: water mains are quite long-lived assets. A company will have many mains systems more than 50 years old. At that time all mains would have been made basically of iron. More recently, other materials such as asbestos cement and latterly uPVC have often been used.)

This model is not one of those that was actually used by any company at AMP2 (which were mostly much more complex), but is one that I developed to illustrate the principles. (It was developed fully in the ABLE software manual – see Section 8.8 below.)

Some terminology: the model is known for obvious reasons as the 'cost model'; quantities appearing on the left-hand sides of equations are called derived quantities; all other qualities in the cost model are called base quantities (BQs). The BQs in this model are Rehab UC$_i$, Iron5$_i$, Iron4$_i$, Iron3$_i$, Decay$_i$, Lowprops$_i$, PresCost$_i$, GrowCost$_i$, Growth$_i$, NewDemd$_i$, LeakRed$_i$, NewProps$_i$, Connect$_i$, NewFac$_i$, OnOffC$_i$.

Prior information will therefore be elicited about the BQs. Data will arrive in the form of engineering study estimates of BQs, and the BLE will be applied to derive new (posterior) means, variances and covariances of BQs. The function of the cost model then was to allow that information to be used to derive also posterior means, variances and covariances of any derived quantities which might be of interest.

A second important consequence of the change to inference about BQs, instead of zone costs, should be mentioned here. It was now necessary to think more carefully about the prior information. The original method had a certain robustness by virtue of working at a highly aggregated level. Errors in specifying prior information about a single quantity would be unlikely to make much difference to the implied information about zone costs. But with BLE updating applied at the level of BQs, misspecification of the prior might produce exaggerated updating, leading to wild posterior estimates of BQs which might have a marked effect on estimates of derived quantities. Section 8.10 sets out the more elaborate elicitation process I developed to try to obtain reliable elicitation, and Section 8.11 relates just how bad some of the elicitation nevertheless proved to be!

## 8.7 Systems of expert assessments

The new approach required companies to furnish engineering study estimates of individual BQs. It was clear that they did not now need to consider studying everything about a zone in order to get a whole zone cost estimate. In fact, as soon as I related this to the engineers it became clear that they might now contemplate doing different kinds of study, in varying depths.

Companies might wish to focus their studies on the need for individual areas of investment, such as a study to examine just the problems of sewer flooding in a catchment, or even on individual assets, like a study of a water treatment works. Furthermore, discussions revealed that engineers often had the option of doing more or less detailed studies of some kinds of problem. For instance, a study to learn about the length of water main in a zone that needed renovation because its internal corrosion was affecting water quality might be a major undertaking. It could involve building accurate network models, sampling water at various points in the network, and even digging out sample sections of pipe. Alternatively, a less intensive study might be based on customer questionnaires. Customers experiencing poor water could then be plotted on a map of the system, and an expert asked to mark the pipes he thought most likely to have led to the observed pattern of problems.

In addition to this complication, there was another important issue which I felt had not been fully resolved in the original approach, and which now became more urgent. This was the matter of treating the study data as posterior estimates – I wanted to have a fully coherent way of doing that.

To tackle these two problems together, I talked with Professor Michael Goldstein of Durham University. Michael is the leading expert on Bayes linear methods, having developed most of the theory himself. (For more details and further references to his many contributions, see Chapter 4 in this book.) The result of our collaboration is Goldstein and O'Hagan (1996). Parts of that paper are highly abstract because one goal of the paper is to provide a thorough and rigorous justification of the structure which is implied by the requirement that a study estimate should replace the prior estimate completely. An outline of that structure can, however, be given in quite simple terms.

Take the example of water quality. The company has $N$ water distribution zones. Let the true length of main to be renovated for water quality reasons in zone $i$ be $X_i$. Now consider three kinds of estimate that can be made of $X_i$. In increasing order of accuracy, there is $X_{1i}$, a prior estimate; $X_{2i}$, an estimate based on customer survey; and $X_{3i}$, an estimate based on more detailed investigation. We suppose that whatever is the best of these estimates that is available for $X_i$ will (in the absence of information from other zones) incorporate and replace any less accurate estimates. So, if the company begins with the prior estimate $X_{1i}$, and learns the intermediate estimate $X_{2i}$, then $X_{2i}$ is the new estimate and not some Bayesian synthesis of $X_{1i}$ and $X_{2i}$. And if they then learn the more detailed estimate $X_{3i}$, it will completely replace the other two as *the* best estimate. It is proved in Goldstein and O'Hagan (1996) that the $X_{ji}$s can then be represented in the form

$$X_{3i} = X_i + E_{3i}$$

$$X_{2i} = X_{3i} + E_{2i} = X_i + E_{2i} + E_{3i}$$

$$X_{1i} = X_{2i} + E_{1i} = X_i + E_{1i} + E_{2i} + E_{3i}$$

where $E_{1i}$, $E_{2i}$ and $E_{3i}$ are *independent* 'errors' with zero prior expectations. This model shows that the less accurate estimates are less accurate *because* they are subject to the addition of more independent error terms. We can see now how each estimate supplants any less accurate estimates. For if we know $X_{1i}$ and $X_{2i}$ then this is equivalent to knowing $X_{2i}$ and $E_{1i}$. Now

$$X_i = X_{2i} - E_{2i} - E_{3i}$$

so to estimate $X_i$ we should subtract estimates of $E_{2i}$ and $E_{3i}$ from $X_{2i}$. But $E_{1i}$ is independent of $E_{2i}$ and $E_{3i}$, and so provides no further information. In the absence of information about $E_{2i}$ and $E_{3i}$ from other zones, we can only estimate them by their prior estimates, i.e. zero, so $X_{2i}$ is our best estimate of $X_i$.

We can also see that information from other zones can be relevant. If in zone $k$ we have both study estimates $X_{2k}$ and $X_{3k}$, then we know $E_{2k}$, and this may provide information about $E_{2i}$. The model does *not* say that $E_{2i}$ and $E_{2k}$ should be independent, and indeed they should be correlated. This is because if the company obtains an estimate $X_{3k}$ in zone $k$ from the most detailed study, and this indicates that the customer survey estimate $X_{2k}$ was too high ($E_{2k} > 0$), then it suggests that the customer survey estimates elsewhere might also tend to be too high.

## 8.8 Three layers of modelling

Apart from the changes described in Sections 8.6 and 8.7, the methodology I proposed to use for AMP2 had all the essential features of the earlier approach. In particular, there would still be all the structuring and simplification needed to make it practical to elicit prior beliefs about the very large number of quantities in the model. So it was necessary first to consider whether independence could be assumed between different quantities in the cost model. With the changed role of this model, it was no longer possible to make quite such sweeping independence assumptions. In the example model (eqs 8.1–8.10), for instance, one could not realistically assume that Iron5$_i$ and Iron4$_i$ are independent. If more grade 5 (the worst condition) mains were found in a zone, the company would expect to find more grade 4 (still unacceptably poor) mains, reflecting a higher level of decay, or poor management, in that zone than expected. So it was sometimes necessary to group BQs together before independence could be assumed between the groups. Then further structuring such as exchangeability (possibly within a stratification) was necessary to facilitate eliciting the variances and covariances within a correlated group of BQs. Remember that each named quantity in the model typically stands for many individual quantities. Iron5$_i$, for instance, stands for the list of lengths of grade 5 iron mains in each zone. Quantities are always assumed to be correlated within each named list. A group may therefore comprise *one or more* such lists.

At this stage it is useful to view the modelling framework as involving three distinct layers of models:

1. The cost model expresses derived quantities of interest in terms of BQs (which might of course also be of interest in their own right). The BQs

are collected into correlated groups, such that independence is assumed between groups. The analysis can then proceed to handle each group quite separately. They are only brought together in any sense when reporting posterior estimates for a derived quantity, when the cost model is used to combine posterior means, variances and covariances for BQs involved in the definition of the derived quantity, which will typically involve drawing posterior information from different groups.

2. For each group, the company specifies the various kinds of estimate that can be made for quantities in that group, ranked into 'levels of uncertainty'. The model structure described in Section 8.7, and more fully in Goldstein and O'Hagan (1996), then represents all these estimates in terms of uncorrelated systems of 'errors'. This is called the 'data model'. It has essentially the same form for all groups, varying only in the numbers of 'levels of uncertainty' specified for each group.

3. The third layer of modelling structures each group by representations such as exchangeability, which is the final step to make elicitation feasible. This is called the 'parametric model' layer.

I will now briefly describe this third layer of modelling; more detail is given in Goldstein and O'Hagan (1997). Let $y_1, y_2, \ldots, y_n$ be the values of a set of correlated 'errors' (i.e. differences between estimates at two adjacent levels of uncertainty) in a group of $n$ quantities. The exchangeability model $y_i = \mu + e_i$ described in Section 8.4 is familiar to statisticians as the simplest case of the general linear model

$$y = X\beta + e$$

where $y$ is the vector $(y_1, y_2, \ldots, y_n)'$, $X$ is an $n \times p$ *design matrix* of fixed coefficients, $\beta$ is a $p \times 1$ vector of *parameters* and $e$ is a vector of *residuals*. The residuals are assumed to be independent of each other and of $\beta$. Then if their variance-covariance matrix is $\text{Var}(e) = D$, a diagonal matrix, and $\text{Var}(\beta) = V$, we have

$$\text{Var}(y) = XVX' + D$$

and so to specify the $n \times n$ matrix $\text{Var}(y)$, which we require in order to apply the BLE theory, we need only elicit the much smaller $p \times p$ matrix $V$ and the $n$ diagonal elements of $D$.

Complete exchangeability is the case where $p = 1$, $X$ is the $n \times 1$ vector of ones and the diagonal elements of $D$ are all equal, so that only two variances need to be elicited. Stratification corresponds simply to the familiar model for one-way analysis of variance. In formulating the general parametric model form, I envisaged allowing much more flexibility in representing engineers' prior knowledge, and so avoiding making such sweeping assumptions as were implied by exchangeability (with or without stratification). In the event, the companies nevertheless stuck very largely to the simple structures for AMP2. The only extra complexity that was used was to consider cross-stratification, which corresponds to a two-way analysis of variance model without interactions.

For practical elicitation, the fact that this model expresses prior uncertainty only very indirectly was a further problem to be solved. I did not expect to be able to elicit beliefs from the engineers accurately by asking them complicated questions about differences between different kinds of estimates of specific quantities. The actual elicitation process is outlined in Section 8.10. For details of the concept of divisible models, which allowed elicitation to proceed at a more natural level, and for a formal explanation of how this modelling fitted together and how the BLE was applied to obtain posterior inferences, the reader is referred to Goldstein and O'Hagan (1997).

Fig. 8.1 expresses diagrammatically the three layers of models. The cost model expresses derived quantities in terms of base quantities. The data

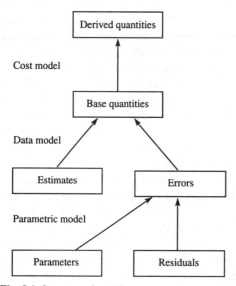

**Fig. 8.1** Structure of models

model expresses BQs in terms of the possible estimates and the errors arising at each level of uncertainty. The parametric model finally expresses errors in terms of parameters and residuals. The diagram also shows the flow of information (means, variances, covariances) about base quantities when we need to report inferences about derived quantities. A fuller description of Fig. 8.1 and the processes involved in prior specification and updating may be found in O'Hagan (1994b).

## 8.9 ABLE

By the time I had developed the theoretical machinery outlined in the last few sections, several more water companies had approached me wishing to use my Bayesian approach for their AMP2. It was clear that the computational effort involved in applying the methodology in even a single company could become enormous. I had developed the software *ad hoc* for the original AMP developments for the two companies I was working for then. I program in APL, which is ideal for prototyping (and a joy to use – some readers may have heard me wax lyrical about APL on other occasions, so I will curb my enthusiasm here!). I developed a suite of APL functions employing ideas of symbolic computation (see O'Hagan, 1990) to handle the four quite different models I was using then within a single general piece of software, but even this kind of approach would be impractical for AMP2. And my time for programming was going to be severely restricted with the number of clients I was contracted to work for.

Although the water companies were supposed to be in fierce competition in 'the marketplace', that was always a figment of an imagination filled with Conservative party dogma. There was no difficulty at all in persuading my clients to club together to fund the commercial development of a software package for AMP2. The initiative and the software needed a name, and I called it ABLE. The name does not really stand for anything, but it incorporates the letters BLE which the water industry had come to identify with my approach, and I thought it conveyed a general air of quality and competence! The sponsors of ABLE comprised Anglian Water, North West Water, Southern Water, South West Water, Thames Water, Yorkshire Water, Bristol Water, Three Valleys Water and the Water Service of the Department of the Environment, Northern Ireland.

The sponsors did not sit passively and just provide money on request. They were involved throughout via a steering committee of representatives of the companies, and had a strong influence on what features ABLE finally provided. (To be strictly accurate, Bristol Water counted as half a sponsor, and did not have a seat on this steering committee.) ABLE was programmed (in APL, first because APL is wonderful, but second because I needed to interact directly with the programmers) by Dyadic Systems Ltd. The overall cost of the project was in excess of £200 000.

The development of ABLE took more than a year, and the companies were concurrently trying to collect the data they needed for AMP2. I was specifying each module as fast as I could, trying to have facilities available as needed. In the event, my technique and specification for designing programmes of studies (i.e. selecting which kinds of studies to apply to which quantities in which zones) were never used because they were not ready in time. The selection of studies had to be done very early because the studies could take months to complete, and companies actually made these decisions based on intuition and expediency. The interested reader can, however, find details of the ABLE optimal design approach in O'Hagan (1994a) and in the ABLE manual (Golbey and O'Hagan, 1994).

## 8.10 Elicitation

An important part of ABLE, and the part which the users found most difficult, is the module concerned with eliciting variance judgements. Elicitation is an area of practical Bayesian statistics which I believe has been given far too little attention. The literature is sparse and much is concerned with individual probabilities rather than eliciting a more complex quantity like a variance. It gives very little guidance for a problem as complex as this. The approach I developed is set out more fully in O'Hagan (1998), with some discussion of other kinds of elicitation.

Consider eliciting the variance of a single quantity $X$, and suppose for the moment that the subject's beliefs about $X$ could be well approximated by a normal distribution. Suppose also that the subject has already stated an estimate $m$ of $X$, which we can take to be the mean. In order to elicit the variance now, it is enough to elicit an interval $(m - a, m + a)$ in which a fixed probability $p$ lies. If, for instance, $p = 0.95$, then the elicited standard deviation is $a/1.96$. However, there is considerable evidence that people are not good at judging 95 per cent probability intervals, even after some training in assessing probabilities. They err consistently on the side of overconfidence, with the result that the inferred standard deviation is too small. On the other hand, some experience I had had suggested that the converse could be true when people are asked for 50 per cent intervals. I therefore decided that I would ask the engineers for 65 per cent intervals. This was easy to explain in terms of asking them to state an interval such that they were twice as sure that $X$ would lie in the interval than that it would be outside the interval. Equivalently, they should feel that odds of 2 to 1 on it lying inside the interval would be fair. One could then simply (as a good approximation) equate the standard deviation to $a$.

The assumption of normality is of course quite strong. Although my discussions with water company engineers suggested that it would be reasonable some of the time, their beliefs about many quantities would be more accurately represented by a positively skew distribution. ABLE was therefore written to accommodate either normal or lognormal distributions. (This is obviously an area for future improvement and introducing the possibility of more kinds of distributions, but the subject then needs to be able to distinguish between the various distributional shapes in some way.) The subject was asked to give a 65 per cent interval without being required to place the limits symmetrically around the estimate $m$. If the subjects' limits were sufficiently asymmetric then ABLE concluded that a lognormal distribution was preferable to a normal distribution, and computed an implied standard deviation by an appropriate formula.

It is generally advisable to check elicitations by a process of overfitting or feedback (see O'Hagan, 1994c). After the subject has stated a 65 per cent interval, ABLE then feeds back two intervals which are implied by the calculated standard deviation (and the normal or lognormal distribution as appropriate) – a 50 per cent interval and an 'almost sure' (99 per cent) interval. If the subject is unhappy with either of these, he or she is asked to revise the stated 65 per cent interval.

This is the process adopted for eliciting a single variance, although I should stress that I am sure there is considerable scope for improvement in future. The user must, of course, specify more than one variance. Consider the simplest case where the parametric model for the groups is the simple exchangeability model described in Section 8.4. To elicit the prior variances and covariances of the base quantities in this case requires the user to specify two variance components. The ABLE elicitation model first asks the user to think about the average of all the quantities in the group (which will typically be a large number). The user places 65 per cent limits on how far the true average might be from the average of the prior estimates, and thereby (after receiving feedback on 50 per cent and 'almost sure' limits) elicits the variance component $v$. The program then asks the user to set 65 per cent limits to say how far the true value for an individual BQ might be from its prior estimate *on the assumption* that the average of the prior estimates is exactly right. After considering feedback again. the user's limits specify the second variance component $w$. Now the variance of an individual prior estimate is $v + w$, when we allow for a systematic error in the prior estimates as well as random (residual) errors. ABLE now provides further feedback in the form of 50 per cent, 65 per cent and 'almost sure' limits for the user to consider, representing limits for an individual BQ *without* assuming that the prior estimates are exactly right on average. If dissatisfied with any of these, the user should go back and modify either the elicitation of $v$ or that of $w$.

However. we should now remember that the prior estimates are usually just one of at least two kinds of estimate (levels of uncertainty) available for BQs in the group, and this process is repeated for each level of uncertainty. Further details of this, and of elicitation under more complex parametric models, is given in O'Hagan (1998).

It should be clear that the elicitation process for any single group of BQs may be far from simple. The water companies may have had dozens of such groups to elicit for, and it is fair to say that they found it difficult! Although I have had extensive discussions with engineers and managers in the various companies, I have to confess that I do not know how to make the task easier. They in turn appreciate that the variances they were being asked to think about are necessary for the proper working of the method, not just in an abstract sense – they have some appreciation of the role that each one plays in the analysis. I took quite a lot of trouble to explain this to them, in non-technical terms, because without their full cooperation and commitment I do not believe that they would have taken the task sufficiently seriously. In fact, they were prepared to spend hours thinking carefully about the inherent inaccuracies, both systematic and random, of all the many estimation processes. And, furthermore, I am sure that this in itself was valuable, because they were not previously in the habit of questioning the accuracy of engineering studies.

## 8.11 Validation

Having said how hard and carefully the water company personnel thought about the elicitation, I must also report that the arrival of estimates in due course, from the programmes of studies, very often proved the elicitation to have been wildly inaccurate!

When a set of study estimates are input, ABLE carries out a number of tests, known as validation tests. Essentially, these consist of comparing the incoming estimates with the prior means, variances and covariances which are implied for those estimates by the elicited prior information. If the new estimates are sufficiently far from their prior expectations, relative to their variance-covariance structure, ABLE flags a warning. At the most obvious level, for instance, ABLE looks at each estimate individually, and if the ratio of the squared difference between the estimate and its prior mean, divided by its prior variance, is too large then a flag is raised. ABLE actually calculates many more kinds of tests (see O'HGB; Golbey and O'Hagan, 1994).

Any such flag is a warning of a possible inconsistency between the elicited prior information and the new data. In general it suggests that the elicited prior variances are too small, although when we remember that the elicitation was conducted in terms of variance components related to a parametric model, it can be difficult to identify where the problem lies. Users were advised to consult me or my colleagues in the event that they obtained validation flags from ABLE. Because of the effort that they and I had expended on careful elicitation I did not expect very many problems, but I was wrong. The water companies experienced tremendous difficulty with this validation process, and subsequent analysis revealed several distinct causes.

The first was simple overoptimism, either in terms of believing that the prior estimation process was much more accurate on average that it proved to be (and hence underestimating systematic error) or in terms of expecting far less variation between zones than was subsequently found (and so underestimating random error). Despite our efforts, this problem arose frequently. The whole exercise was very illuminating for the companies, in both the above respects. They had seriously believed that their prior information was unrealistically good, and were very surprised at the extent of zone to zone variation in quantities they thought would be much more homogeneous.

A second cause was outliers. In many cases, their elicited variances were quite accurate for values of a BQ in most zones, but some zones would produce wildly different values, for all kinds of special reasons. Here was an example of how strict use of Bayes linear methods could be inappropriate. For an exchangeable base quantity, posterior estimates of values of the BQ in unstudied zones depend on the mean of the study estimates. Outliers will exert an unreasonably strong influence on such estimates, and one would prefer to use some kind of robust method which reduced the weight given to extreme observations. In effect, Bayes linear methods are most appropriate when the joint distribution of the relevant random variables is approximately normal, whereas the occurrence of outliers suggests heavy-tailed distributions. It was too late, when these problems arose, to develop new theory and I simply advised the companies to use special features of ABLE to downweight extreme observations in an *ad hoc* way. It was far from being a satisfactory solution, and this is an area where work is still ongoing.

The third principal cause of validation problems lay in the simplifying structural assumptions we made. As discussed in Section 8.4, the engineers often wished to assess their uncertainties in terms of percentage errors, so that in many BQs the elicited standard deviations were expressed as proportional to their prior estimates. This seemed quite innocuous at the time, but so often it is the least considered assumptions that cause the problems! This format was naturally used for BQs where prior estimates varied substantially from zone to zone, and without the users really being aware of the implication, zones with small estimates were given correspondingly small standard deviations. I had become aware of this consequence relatively early in the exercise when instances arose of zero estimates, which were thereby given zero variances. I warned the engineers that this was only proper if a zero estimate meant absolute certainty that the quantity would be zero in that zone, and advised them to beware of small estimates being given unrealistically low variances. However, in the pressure of the massive task which constructing the AMP was proving to be, the warnings were mainly unheeded and serious validation problems followed. Another area of current research is to provide simple variance elicitation procedures for more realistic assumptions about how the prior uncertainty should vary with the magnitude of the prior estimate.

## 8.12   Lessons learned

This has been a sustained collaboration over several years, which at times has taken a sizeable part of my time and energy. I am already beginning to address various perceived deficiencies and to make important improvements to the methodology in preparation for AMP3 (expected in 1999). Such a collaboration could not succeed unless it is clearly to the benefit of both sides. For me, this is an exciting opportunity to apply Bayesian methods (to which my commitment is well known!) in novel ways, and to see the Bayesian approach succeed and spread. As the references to this paper show, it has given me a string of publications which testify to the many interesting and challenging problems it has raised. I do, of course, get paid as a consultant. This was initially conducted at Warwick University through the Statistical Consultancy Unit, and now through Nottingham University Consultants Ltd. An important feature of such arrangements is that a proportion of the fees go to the university and to my department, providing a nice 'pot of gold' in times of general financial stringency for universities.

The water companies obviously benefited too. A variety of advantages that they found in the Bayesian approach have been listed in O'Hagan and

Wells (1993). Rather than speculate in great depth here. I will just mention one factor which I believe is an asset for Bayesian statistics generally. A Bayesian analysis must ask about prior information. Whereas the same underlying statistical model might apply in a variety of applications, and so attract the same classical analysis in each case, prior information makes each application unique to a Bayesian. Of course this fact works against Bayesian statistics by making it difficult to develop general purpose software, but in serious applications it is a great asset. The Bayesian must take an interest in the context of the problem and the client's prior knowledge. The Bayesian must take pains to model the prior information accurately and must be sensitive to precisely what form of inferences the client requires. Not only is this actually really interesting for the Bayesian statistician, it also helps to build a relationship of trust between statistician and client. Importantly, the client very often comes to understand his own problem better. I believe that the water company engineers and planners have come to understand much more clearly the uncertainties inherent in trying to estimate the various costs and commitments that go to make up the AMP than they would have achieved through any classical analysis.

Any practical application of statistical methods must make assumptions, and almost inevitably the failures of some of these come back to haunt the statistician. This project was no exception, as the discussion of assuming standard deviations proportional to means in the last section shows. But an intriguing message to me from this collaboration concerns the unconscious assumptions that we bring to a problem by virtue of our own previous experiences and training. All my training as a Bayesian statistician had taught me that one uses Bayes' theorem to synthesize prior information and data, so when I proceeded to do this I was completely unprepared for finding that this was not what the client wanted. As discussed in Section 8.5, the data are in effect the posterior estimates. I believe that this feature might actually hold much more widely, and it has been, to me, a fascinating and enduring problem to determine how best to model such a situation.

I hope the reader will find other lessons to learn in this chapter. It has been for me a story like a good thriller, with unexpected twists in the plot and with a momentum that carries the reader along enthralled. As the publisher's blurb might say, it has been 'unputdownable'.

# References

Ericson, W.A. 1969: Subjective Bayesian models in sampling finite populations (with discussion). *Journal of the Royal Statistical Society, Series B* **31**, 195–233.

Golbey, L.A. and O'Hagan, A. 1994: *ABLE, version 1*. Nottingham University Consultants Ltd, Nottingham University.

Goldstein, M. and O'Hagan, A. 1996: Bayes linear sufficiency and systems of expert posterior assessments. *Journal of the Royal Statistical Society, Series B* **58**, 301–16.

Goldstein, M. and O'Hagan, A. 1997: Modelling diagnostics for systems of expert posterior assessments. (In preparation.)

O'Hagan, A. 1984: Bayes linear estimators for finite populations. *Statistics Research Report* **58**, University of Warwick.

O'Hagan, A. 1987: Bayes linear estimators for randomized response models. *Journal of the American Statistical Association*, **83**, 503–8.

O'Hagan, A. 1988: *Probability: Methods and Measurement*. London: Chapman and Hall.

O'Hagan, A. 1990: Variance of an arithmetic expression – an example of symbolic computation and recursion. *Vector* **6**(3), 80–7.

O'Hagan, A. 1994a: Bayesian methods in asset management. In Barnett, V. and Turkman, K.F. (eds) *Statistics for the environment, water-related issues*. Chichester: Wiley, 235–47.

O'Hagan, A. 1994b: Robust modelling for asset management. *Journal of Statistical Planning and Inference* **40**, 245–59.

O'Hagan, A. 1994c: *Bayesian inference*. Volume 2B of Kendall's Advanced Theory of Statistics. London: Edward Arnold.

O'Hagan, A. 1998: Eliciting expert beliefs in substantial practical applications (with discussion). *The Statistician* **47** (in press).

O'Hagan, A., Glennie, E.B. and Beardsall, R.E. 1992: Subjective modelling and Bayes linear estimation in the UK water industry. *Applied Statistics* **41**, 563–77.

O'Hagan, A. and Wells, F.S. 1993: Use of prior information to estimate costs in a sewerage operation. In Gatsonis, C., Hodges, J.S., Kass, R.E. and Singpurwalla, N.D. (eds) *Case studies in Bayesian statistics*. New York: Springer-Verlag, 118–63.

Smouse, E.P. 1984: A note on Bayesian least squares inference for finite population models. *Journal of the American Statistical Association* **79**, 390–2.

Sugden, R.A. and Smith, T.M.F. 1984: Ignorable and informative designs in survey sampling inference *Biometrika* **71**, 495–506.

# 9 The rise and fall of the US Department of Energy's Environmental Restoration Priority System[1]

M W Merkhofer, K E Jenni,
C Williams

## 9.1 Introduction

The United States Department of Energy (DOE) is engaged in the largest civil works project ever undertaken – a 30-year, $100 billion clean-up of hazardous waste sites (National Research Council, 1994a). The sites, contaminated with radioactivity and toxic chemicals, are a legacy of 50 years of nuclear weapons production. Over $7.5 billion has been spent on clean-up since 1989 (US Department of Energy (hereafter USDOE), 1994). Concerns regarding the risks posed by the sites, the costs of the effort and the slow rate of progress are routinely reported in the media.

To help manage the clean-up, DOE's Office of Environmental Restoration (ER), within the Office of Environmental Management (EM), developed a sophisticated decision support system called the Environmental Restoration Priority System (ERPS). ERPS made recommendations for how to divide the annual ER budget among programmes and sites based on considerations of site-specific health risks, regulatory requirements, the costs and effectiveness of proposed clean-up activities, and other relevant considerations. The system was praised in technical peer review and appeared to improve the quality of DOE's funding allocation decisions. However, ERPS became a focal point for criticism from parties outside the DOE. In September 1993,

[1] The authors wish to thank individuals involved with ERPS who reviewed the discussion in this chapter and provided helpful comments and suggestions, including Frank W. Baxter, Thomas A. Cotton, John E. Kelly, and, especially, Thomas P. Longo.

the assistant secretary for EM announced that 'serious problems' had been identified and that the system had been 'deferred'.[2]

This chapter expands upon a previously published paper on ERPS (Jenni *et al.*, 1995). The authors were involved in the design, application and public participation aspects of the system. Our objective is to describe the history of the system, including its motivation, how it worked and the controversy in which it became embroiled. Lessons from ERPS are particularly germane, given increasing government interest in risk-based prioritization (Carnegie Commission on Science, Technology, and Government, 1993; US Environmental Protection Agency (hereafter USEPA), 1990a, 1993).

## 9.2   The DOE decision problem

Managing the financial aspects of the clean-up of contaminated sites requires DOE to make two kinds of decisions: (1) how much money should be requested annually from Congress?; and (2) how should appropriated funds be allocated across competing needs? Funding requests, including proposed activities and costs, must be submitted about 2 years before allocations are actually made. The funds ultimately provided are typically insufficient to allow full funding for all desired activities. DOE must, therefore, determine how to divide the annual budget among DOE's ER programmes, field offices and the more than 100 facilities responsible for showing progress at local sites (Fig. 9.1). Also, within a given facility, the funds must be divided

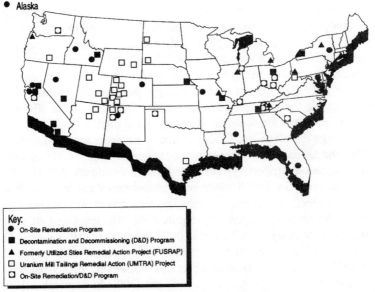

**Key:**
- ● On-Site Remediation Program
- ■ Decontamination and Decommissioning (D&D) Program
- ▲ Formerly Utilized Sties Remedial Action Project (FUSRAP)
- ☐ Uranium Mill Tailings Remedial Action (UMTRA) Project
- ☐ On-Site Remediation/D&D Program

**Fig. 9.1**  DOE facilities and programmes evaluated using ERPS

[2] 'Questions from Senator Johnston', written responses from T. Grumbly prepared in September 1993 following testimony on July 29, 1993, to the Committee on Energy and Natural Resources, United States Senate, p. 29.

among the various clean-up activities that the facility's ER managers would like to conduct.

Much of the clean-up work proposed for sites is in direct response to legal mandates. Applicable laws and regulations include the Comprehensive Environmental Response, Compensation, and Liability Act; the Superfund Amendments and Reauthorization Act; the National Environmental Policy Act; the Atomic Energy Act; the Federal Facilities Compliance Act; and numerous individual installation consent orders and agreements, Federal facilities agreements and operating permits. Despite this detailed regulatory structure, funding decisions cannot be based on legal mandates alone. The budget allotted by Congress may not be sufficient for all required work. Even if it is, not all sites are adequately covered by requirements, and, for those that are, the amount of money needed to achieve an appropriate level of confidence that mandates will be met is uncertain. Comparing funding needs is difficult. Little hard data exist on the adverse impacts of the sites on local communities or on the effectiveness of proposed clean-up activities. Risk estimates, for example, vary in quality, but are generally highly uncertain.

Because of these difficulties, funding choices must ultimately be based on judgment. Obtaining unbiased judgments and incorporating such judgments into the decision making process is complicated by the decentralization of information and management responsibilities and by differences in incentives at different levels within DOE. DOE headquarters (HQ) has responsibility for preparing budgets, but detailed knowledge regarding specific clean-up efforts is held by field and facility personnel, local regulators and contractors. Contractors want to ensure adequate funding for their work, and field and community representatives want to maintain employment in the face of declining defence mission budgets. Neither contractors nor DOE field personnel want to risk civil or criminal liability for failing to meet legally mandated schedules and milestones. Meanwhile, HQ has responsibility for keeping total requests in line with budget realities. Information about environmental problems and proposed activities, as well as the justification for those activities, must be communicated from the facilities to HQ, and, in turn, from HQ ultimately to Congress. Information about priorities and budget constraints must be communicated from HQ to the facilities. Compounding these decision making difficulties are widespread fear of hazardous waste and public distrust of the agency.

Recognizing these complexities of the clean-up and budgeting process and the importance of using available funds wisely, numerous parties have called on DOE to establish some formal approach to setting priorities (Committee on Armed Services, 1987; National Governors' Association and the National Association of Attorneys General, 1990).[3] ERPS was developed to respond to this need.

---

[3] 'Defense waste cleanup: a proposal for a national solution', transmitted to J.T. Watkins, Secretary of Energy, from governors of 10 states in letter dated April 14, 1989.

## 9.3   Description of ERPS

ERPS was designed to collect and document information relevant to budgeting decisions and to use that information to help identify funding allocations that maximize the achievement of ER objectives. It consists of an analytic model plus a systematic process and associated tools for collecting the inputs needed to run the model.

### 9.3.1   Technical components

The ERPS model consists of three technical components (Voth *et al.*, 1993). First, at the heart of ERPS lies a multi-attribute utility (MAU) model (Keeney and Raiffa, 1976) for evaluating the benefit of clean-up activities. The model quantifies benefit using a utility function, which was formally elicited from policy level DOE managers and reflects the objectives and values underlying the ER programme. The utility function has an additive form (verified using independence checks) and accounts for six types of benefits: (1) reduced health risks to workers and the public; (2) reduced adverse impacts to the natural environment; (3) reduced adverse socio-economic impacts; (4) compliance with applicable laws and regulatory requirements; (5) reduced ultimate costs of clean-up (e.g. by cleaning up a spreading problem early); and (6) reduced uncertainties related to risks and costs. Fig. 9.2 identifies these criteria and the associated subcriteria included within the MAU model. Table 9.1 summarizes the weights assigned to the criteria and some of the underlying value judgments implied by the weights.

The second technical component of ERPS is a decision-analytic, value-of-information calculation for estimating the benefits of reducing uncertainties. A separate component of the system is devoted to this benefit criterion because much ER funding is spent on studies (e.g. site characterization studies), and because, in the absence of computational aids, estimating the value of reducing uncertainties is difficult. Value-of-information theory (Howard, 1966) is used to compute the economic value of completely resolving risk and cost uncertainties. This resulting 'value of perfect information' is then scaled according to the judged ability of the studies to reduce uncertainty.

The third ERPS technical component is a combinatorial optimization routine for calculating efficient allocations of funds across facilities. As clarified below, applications of ERPS estimated the benefits of alternative funding levels for each facility. The optimization is a branch-and-bound approach for determining those combinations of facility funding levels that are *efficient* in the sense that no other funding options exist that lead to the same or lower total cost but greater total benefit. Since the system considers about 100 facilities and, on average, five alternative funding levels for each facility, roughly $10^{10}$ combinations of funding must be considered. The optimization routine sorts through the possibilities in an efficient manner.

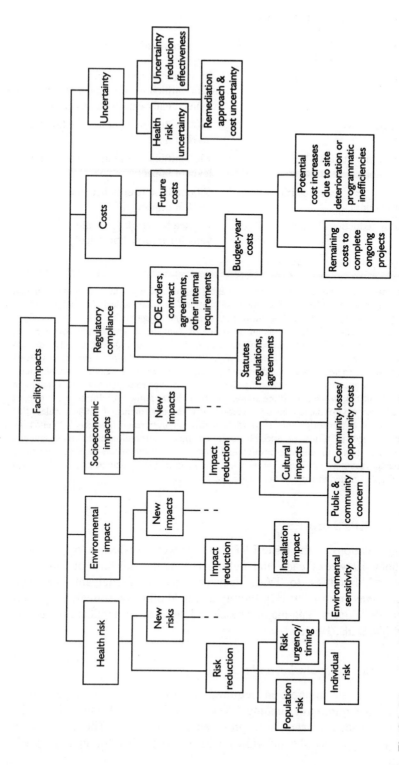

**Fig. 9.2** Criteria for estimating benefits and costs. Note: dashed lines below 'new risks' and 'new impacts' indicate structures similar to those shown under 'risk reduction' and 'impact reduction', respectively

**Table 9.1**   Priority system weights

| Criterion | Relative weight | Value trade-off judgment[a] |
|---|---|---|
| Health risk | 36% | $5 million per health effect avoided $200 million to eliminate a $10^{-1}$/yr risk to maximally exposed individual |
| Environmental risk | 13% | $400 million to eliminate the highest level of impact |
| Socio-economic impact | 9.5% | $300 million to eliminate the highest level of impact |
| Regulatory responsiveness | 9.5% | $300 million to eliminate the highest level of impact |
| Uncertainty reduction | 32% | Implied by weight on health risk and value-of-information calculation |
| Total | 100% | |

[a] The weights were provided by senior managers within DOE's Office of Environmental Restoration in a series of formal, facilitated 'elicitations' in which a decision analyst derived the managers' willingness to trade achievement along one objective for achievements along others. A 'health effect' is defined as an incidence of a major, adverse health consequence, such as premature fatality, severe neurotoxic effect, disabling birth defect, etc. The system scoring instructions (Applied Decision Analysis, 1993) provide definitions for various levels of environmental, socio-economic and regulatory impacts to which the environmental, socio-economic and regulatory value trade-off judgments apply. For example, the value judgment for environment risk implies a willingness-to-pay up to $400 million to eliminate the highest level of adverse environmental impact. According to the environmental impact scales, this level consists of a 'very high threat' to 15 'sensitive environmental resources'. 'Very high threat' is defined in terms of specific toxic chemical and radionuclide environmental concentrations, biota monitoring results and rates of environmental decline. Sensitive environmental resources are identified by a list of 30 entries (e.g. a habitat known to be used by Federally designated threatened or endangered species).

### 9.3.2   The ERPS process

Seven applications of the basic ERPS system were conducted. Applications took about 10 weeks each and culminated in the delivery of numerous reports and briefings to DOE decision makers. Participants included representatives from DOE HQ, facility and field office personnel, and ERPS technical developers. The major phases and steps of an application were as follows (USDOE, 1991a).

#### 9.3.2.1   System preparation and training

Preparing for an application required setting system parameters (e.g. specifying the minimum funding level for each facility) and training the individuals, called *scorers*, who provided system inputs. The scorer for a facility was typically the individual with responsibility for managing ER activities at that facility. Training consisted of 2 day workshops involving lectures and case study exercises. The exercises required scorers to practise developing inputs using information provided for a hypothetical facility. Approximately 60 individuals participated in each training session. Software for collecting and error-checking inputs was distributed to participants at the training sessions.

### 9.3.2.2 Inputs

*Identifying, classifying and ranking activities*   After training, scorers had 3–4 weeks to prepare their preliminary system inputs. First, each facility scorer identified all activities that the facility's ER personnel wished to conduct during the target year (the fiscal year for which the budget was being planned) and obtained cost estimates for each activity. Then, using criteria and a screening process provided in a screening and scoring manual (Applied Decision Analysis, 1993), scorers divided their proposed activities into three classes:

- *Class 1: Emergency activities* – activities needed to address problems that are creating or will be creating significant adverse impacts to public health before work in the target year could take effect.

- *Class 2: Time-critical activities* – activities needed to address problems that must be acted on or studied soon, but not necessarily before the target year. These included activities to stabilize sites that might otherwise deteriorate enough to pose an imminent health threat and activities to study sites that could be posing current health or environmental risks.

- *Class 3: Other high benefit and time sensitive activities* – additional activities that facility personnel wanted to conduct during the target year, including all mandated activities and activities needed to address future potential, rather than existing or immediate, risks.

If any emergency activities were identified, immediate funding was provided for those activities. Thus, emergency activities were, in effect, removed from the system. Two emergency activities were identified during the first system application, but none thereafter. Among the remaining activities, time-critical activities were accorded highest priority. Roughly one in every 100 proposed activities was identified as time-critical. As explained below, full funding was always recommended for time-critical activities.

Scorers next ranked class 3 activities in order of relative priority. No specific ranking process was specified, allowing managers the flexibility to incorporate local considerations. The ranking methods used by scorers ranged from formal, analytical approaches to strict reliance on the schedules and priorities negotiated with local regulators and documented in compliance agreements. A few facilities obtained participation by their local citizen advisory groups in the ranking process.

*Developing and scoring budget cases*   Scorers used the rankings to group activities into budget cases, as illustrated in Fig. 9.3 and Table 9.2. Each case represented the set of activities that would be conducted should the facility receive a specified level of total funding. Between three and 10 budget cases were typically defined for each facility, depending on the magnitude of the requested funds:

- The *maximum case* consisted of all activities that could be accomplished by the facility within the target year, given unlimited funding.

- The *minimum case* was the set of activities that would be conducted for a specified percentage (usually 70 per cent) of the prior year's budget. This represented the minimum level of funding considered for each facility.

- *Intermediate cases* consisted of sets of activities that would be conducted at funding levels between the maximum and minimum cases.

Scorers included time-critical activities in every case. Therefore, all time-critical activities were always recommended for funding by the system.

Each budget case was evaluated and scored according to its estimated ability to achieve DOE ER objectives (Fig. 9.2). Detailed scales were provided for each measure, and scorers used those scales to express their estimates (USEPA, 1990b). Specifically, under each budget case, scorers estimated risk levels and timing, environmental impacts, socio-economic impacts, degree of regulatory compliance, costs and uncertainties. For illustration, Table 9.3 shows the scale used by scorers to select risk timing

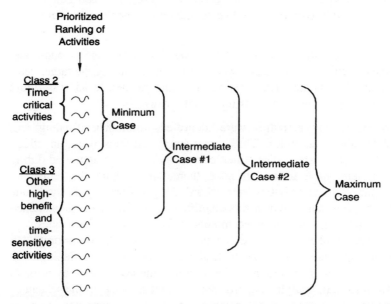

**Fig. 9.3** Schematic illustrating alternative budget cases. The composition and cost of cases were adjusted based on feasibility, dependencies and ability to eliminate or slow down work on specific activities

**Table 9.2** Sample budget cases proposed by facility personnel

| | | | Funding level ($ millions) | | | |
|---|---|---|---|---|---|---|
| Class | Rank | Activity description | Minimum Case | Intermediate no. 1 Case | Intermediate no. 2 Case | Maximum Case |
| 2 | 1 | Look for unidentified contamination source | 1.5 | 1.5 | 1.5 | 1.5 |
| 3 | 2 | Remove barrels | 1.0 | 1.0 | 1.0 | 1.0 |
| 3 | 3 | Begin demolishing building | 0.5 | 1.5 | 1.5 | 1.5 |
| 3 | 4 | Continue risk assessment | 0.0 | 0.0 | 1.0 | 1.5 |
| 3 | 5 | Remove contaminated soil | 0.0 | 0.0 | 0.0 | 0.5 |
| **Totals** | | | 3.0 | 4.0 | 5.0 | 6.0 |

**Table 9.3** Performance measure scale for scoring health risk timing. To be used to assign a number between 1 and 7 indicating how soon exposures are estimated to occur

| Scale | Time to exposure |
|---|---|
| 1 | No sooner than 100 years from the target year |
| 2 | Approximately 70 to 100 years from the target year |
| 3 | Approximately 50 to 70 years from the target year |
| 4 | Approximately 30 to 50 years from the target year |
| 5 | Approximately 10 to 30 years from the target year |
| 6 | Approximately 5 to 10 years from the target year |
| 7 | Within the 5 year planning horizon (target year through target year +4) |

scores. Scorers typically consulted with risk experts to develop health and environmental scores; cost estimators to develop costs; legal staff to develop regulatory compliance scores; and public affairs staff, community specialists or, in some cases, community representatives to develop socio-economic scores.

The quality of information available to support the scoring process differed dramatically from facility to facility, so scorers were directed to use the best information available for their facility. For example, sites with completed characterization studies used the results of their formal risk assessments to obtain risk scores, while facilities with no formal risk data relied on best professional judgment, considering the quantities and nature of suspected contaminants and potential for human exposures. Most sites had some risk data available, either from a DOE Environmental Survey (USDOE, 1988a) or from applications of EPA's Hazard Ranking System (USEPA, 1990b), a site-ranking model used in the Superfund programme.

### 9.3.2.3   Review and analysis

Quality assurance reviews were held to promote objective and accurate scores. Facility scorers explained and defended their activities, rankings, budget cases and scores to scorers and peers from other facilities, operations office managers and HQ personnel. The reviews included a comparison with scores assigned in earlier applications. For example, if facility scorers had previously estimated that a level of funding would result in significantly lower risk scores, and if that level of funding had been or was projected to be provided, then any deviation of current risk scores from earlier estimates had to be explained in terms of new information or a failure to achieve estimated risk reductions. The review also included inspecting the computed benefit vs. cost curves for each facility. These curves, derived from the MAU model and value-of-information components of the system, typically had a convex shape indicating declining marginal benefits with increasing costs (Fig. 9.4). If the curve had a concave shape, scorers were asked why

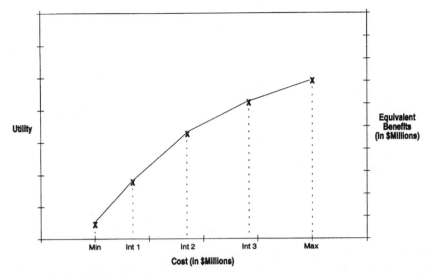

**Fig. 9.4**  Typical facility benefit vs. cost curve

activities with high benefit-to-cost ratios could not be included in lower-cost budget cases.

The finalized scores provided the basis for the optimization routine, which aggregated the facility benefit–cost curves and calculated the allocations producing the greatest total benefit for any specified total cost. The system produced several outputs, including a table of recommended allocations, an aggregated ranked list of activities, and graphs showing how increased or decreased total budgets would affect the estimated health risks, environmental impacts, etc., at various facilities. An important output was a graph showing the relationship between total benefits and total costs (Fig. 9.5). The point on the curve where a dollar of additional cost was estimated to fund activities yielding one dollar of incremental benefit was called the 'one-to-one funding level' – the point with unit slope.

Sensitivity analyses were conducted to investigate the sensitivity of results to assumptions. For example, the impact of regulatory requirements on funding needs was of particular interest. Analyses were conducted both with and without a constraint requiring sufficient funds to permit full regulatory compliance and with weights on regulatory compliance ranging from 0 to 100 per cent. However, applications showed that full regulatory compliance was generally unattainable, even with maximum funding, due to technical or schedule infeasibility. Therefore, when the constraint could not be applied, the system was asked to estimate the funding required to achieve the maximum possible regulatory compliance. Funding requirements and allocations were found to be very sensitive to the weight assigned to regulatory compliance. For example, to achieve maximum possible regulatory compliance, necessary funding levels were typically estimated to be about 50 per cent greater than the funding levels actually chosen. With

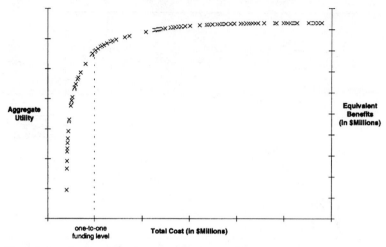

**Fig. 9.5** Sample plot of benefits and costs of optimal allocations

zero weight on regulatory compliance (implying no value to complying with regulations that produce no additional health, environmental, socio-economic, or cost-reducing benefit), one-to-one levels were typically about one-third of actual funding levels.

### *9.3.2.4   Decision making*

At the conclusion of each application, DOE decision makers were briefed on the results. Applications conducted prior to finalizing the ER budget request were used to help decide which cases to include in the request. Applications conducted after the budget was set were used to help determine how to divide the funds among field offices, programmes and facilities (USDOE, 1988a).

## 9.4   The ERPS controversy

The ERPS design evolved in a continuing attempt to satisfy the needs and concerns of stakeholders. The shortcomings that eventually led to its shelving are best understood by reviewing the system's history. Table 9.4 provides a timeline.

### 9.4.1   The Program Optimization System

ERPS evolved from an earlier priority system called the Program Optimization System (POS) (USDOE, 1988b), developed in 1988 in response to a Congressional directive advising DOE 'to establish a priority system ... for the application of funds for environmental restoration' (USEPA, 1993). At that time, DOE's clean-up activities were not organizationally separate from

**Table 9.4** ERPS timeline

| Date | Status |
| --- | --- |
| Spring 1988 | Congress directs DOE to establish a priority system |
| Summer 1988 | Defense Programs (DP) develops the Program Optimization System (POS), predecessor of ERPS |
| November 1988 | Initial application of POS |
| Summer 1989 | DOE forms Office of Environmental Restoration and Waste Management (EM), briefs State and Tribal Government Working Group (STGWG) on plan to develop ERPS, and establishes External Review Group (ERG) |
| October 1989 | ERG meets, expresses doubt about need for ERPS and concern about conflict with legal agreements |
| November 1989 | ERG meets, requests straw man design and suggests adding criteria related to uncertainties and sociocultural concerns |
| February 1990 | ERG issues position statement, including demand for involvement in any application of ERPS |
| Spring 1990 | DOE provides straw man design similar to POS and conducts pilot application without ERG. Individual ERG members interviewed |
| September 1990 | ERG meets, objects to pilot application, suggests ERPS be used only for 'discretionary funding' and expresses concerns about data quality and system complexity. ERPS application to FY92 funding initiated |
| October 1990 | ERG sends letter to DOE, objecting to pilot application and expressing other concerns |
| December 1990 | DOE responds to February position statement, asserts intent to broaden public involvement but cites inability to share pre-decisional budget data |
| January 1991 | ERPS modified to include optional constraint allowing system to consider only funding allocations that meet legal requirements. DOE agrees to work with outside agency to improve risk data quality. ERPS application to FY93 initiated |
| February 1991 | Final ERG meeting. DOE promises public involvement but leaves specifics to field offices. ERG complains that its main concerns were ignored |
| April 1991 | ERPS receives very favourable independent technical peer review |
| April 1991 | ERPS reapplication to FY93 initiated with limited public involvement provided by some field offices |
| July 1991 | ERPS reapplication to FY92 initiated |
| September 1991 | Federal Register Notice requests comments on ERPS |
| October 1991 | DOE holds workshop to encourage FRN comments but declines to present public participation plan. Participants object to display of one-to-one funding level, lack of public involvement plan |
| February 1992 | FY94 application cancelled |
| September 1992 | DOE responds to FRN comments, saying public involvement will be coordinated with other public involvement activities, and reiterates inability to share pre-decisional budget information |
| December 1992 | First meeting of National Academy of Sciences (NAS) review panel |
| February 1993 | Second NAS panel meeting. Keystone report released |
| July 1993 | NAS review cancelled, system shelved |

other missions. Defense Programs (DP) had responsibility for clean-up activities at weapons complex facilities, and it was this office that developed POS.

To ensure a design acceptable to the field personnel who would be required to provide inputs, an advisory group of facility representatives was established. The group was initially sceptical, but admitted that some means was needed to 'level the playing field' and reduce incentives to be a 'squeaky wheel'. Group members insisted that facility ER decisions not be micromanaged by an HQ priority system. The typical priority system design, wherein activities are ranked and then funded from the top down until the budget is exhausted, was very unpopular. As one participant noted, 'There may be fat in the system, but you don't cut fat by lopping off fingers and toes'. It was because of these arguments that POS (and ERPS) was designed to score portfolios of activities (rather than individual activities). The goal was a bottom-up approach where facility managers (and, potentially, local stakeholders), not some centralized system, specified where and how funding shortfalls would be handled and estimated the consequences of those shortfalls. Also important was the concern that activity-ranking systems cannot account for interdependencies among activities and partial funding opportunities.

POS differed from ERPS in three ways. First, POS was a simpler system that lacked activity screening, the value of information analysis, and criteria related to socio-economics. Second, POS applied only to the 16 facilities within DP. Third, POS was intended solely as an internal planning tool. No outside involvement was planned or included.

A first application of POS was conducted in late 1988, 'to help plan the FY90 budget request', and a second 6 months later, 'to provide a basis for finalizing allocations among field offices' (Merkhofer *et al.*, 1989). Scorers who anticipated that the system would be easy to game quickly learned otherwise. For example, padding cases with low benefit work failed because it decreased benefit-to-cost ratios and increased the risk that the facility would receive only minimum funding. The strategy of putting critical work in only high-cost cases was easily spotted from the facility benefit vs. cost curves (Fig. 9.2). Finally, a strategy of claiming the full benefit of multiyear projects with high costs in outyears backfired because such activities generally fared worse than if only annual benefits and costs had been used. Once it was demonstrated that attempts to game the system would be publicly revealed by the QA process, gaming attempts became very rare.

POS applications estimated one-to-one funding levels close to planned levels, but recommended radical changes in the allocation of funds across facilities. The most controversial recommendation was a 50 per cent cut in funding for the Hanford facility, which had been receiving a large share of the ER budget. Most of Hanford's proposed work consisted of studies rather than actual clean-ups. It was suspected that the inability of POS to account for the value of studies to reduce uncertainties resulted in undervaluing the benefits of funding Hanford. Partly because of these recognized weaknesses and partly because the funding changes recommended were so radical, POS applications produced little if any impact on actual budget allocations. Nevertheless, DOE decision makers were strong supporters of the system. It provided a conduit by which HQ obtained detailed documentation from the field regarding clean-up problems and expectations.

### 9.4.2    The DOE reorganization

In July 1989, DOE reorganized and created the Office of Environmental Restoration and Waste Management (later renamed Environmental Management, EM) to oversee and manage all environmental activities at DOE sites. This new office centralized, elevated and organizationally separated site clean-up and waste management activities from other DOE missions. EM activities are coordinated under three offices: Waste Management, which oversees the treatment, storage and disposal of hazardous, radioactive and mixed wastes at currently active sites; Environmental Restoration (ER), which is in charge of containment, characterization, and clean-up activities at inactive or surplus sites; and Technology Development, which oversees research into new clean-up technologies (USDOE, 1993).

The early initiatives of the new EM organization included establishing an annual 5 year planning process and a 'risk-based priority system'. The scope of the priority system was limited to inactive sites under ER. However, if the initial system was successful, the plan called for developing similar systems for the other ER offices. The same design team that had developed POS was assigned responsibility for developing the new system for ER.

### 9.4.3    The External Review Group

The organizational changes taking place at DOE reflected, in part, an effort by senior agency officials to open up the closed decision making processes historically followed by the security-conscious agency (USDOE, 1989). Consistent with this goal, the ERPS design team was advised that 'development and implementation of the system should be accomplished with participation of all interested parties' (Longo *et al.*, 1990). An External Review Group (ERG) was established to participate in system development. ERG membership was largely a subset of an existing external advisory group called the State and Tribal Government Working Group (STGWG), impanelled to review the EM 5 year plan. The ERG included representatives from affected states, Indian Tribes, national-level governmental organiz-ations (the National Governors Association, the National Conference of State Legislators, and the National Association of Attorneys-General), interest groups (the Natural Resources Defense Council and the Environmental Defense Fund) and the EPA.

To provide input for the first ERG meeting, a questionnaire was distrib-uted to STGWG members to solicit opinions regarding desirable characteristics for the priority system. Of eight potential priority system characteristics, 'allows for participation by the states and other interested parties in the development of inputs' was rated by respondents as most important. Eighty-six per cent of respondents indicated this characteristic was 'essential'. The potential characteristic rated as least important was 'simplified ... it is easy for the average citizen to understand'. Only 14 per cent of respondents called this characteristic 'essential'.

In October 1989 the first meeting with the ERG was held. DOE explained the ER funding process, and contractors presented an introduction to formal priority systems. POS was presented as an example of one type of system. The design team identified design decisions for which guidance was sought, including what decision unit should be evaluated (e.g. individual activities vs. a portfolio of activities) and what criteria should be used for the evaluation. However, rather than engage in a discussion about design, the ERG questioned the need for a priority system and chose to use the meeting to work independently towards drafting a consensus statement of ERG concerns and recommendations. Some ERG members were or had been involved in negotiating triparty agreements – the legally binding requirements that specify site-specific priorities, milestones and clean-up schedules. These agreements, many ERG members argued, provided the sole basis for funding decisions.

At the second ERG meeting, in November 1989, the design team continued to press for design guidance. The design team and the ERG did develop a hierarchy of evaluation criteria, similar to Fig. 9.2. This exercise identified several criteria not included within POS, including the desirability of resolving site uncertainties and the importance of addressing sites creating adverse impacts on the cultural or religious practices important to certain groups (e.g. Native Americans prevented from using contaminated land on reservations). The ERG was reluctant, however, to recommend other technical characteristics for the system. Rather than continue this step-by-step design process, the ERG suggested that a more efficient approach would be for the DOE to develop a 'straw man' design – a proposed design that could be reviewed in its entirety and critiqued by the ERG. Key to the concept was that there be no initial DOE commitment to the straw man.

On February 9 1990, the ERG released its consensus statement. It included the declaration, 'The External Review Group (ERG) opposes DOE's unilateral application of any prioritization system'. In other words, the ERG called for involvement of external parties in any system application. The ERG acknowledged that 'a prioritization system may be useful in consensus situations, i.e., negotiating (or renegotiating) schedules', but emphasized that the system must hold to several basic principles, including: (1) imminent and substantial threats to human health or the environment should be given highest priority; (2) DOE must comply fully with all legally binding commitments; (3) the nature and extent of releases and impacts must be characterized as expeditiously as possible; (4) Tribal sovereignty and rights must be recognized; and (5) all concerned parties must participate fully in the implementation of the DOE priority system. Finally, the ERG emphasized, 'budgetary constraints do not affect DOE's binding legal obligations'.[4]

### 9.4.4   The decision to conduct an application to FY92 budgeting

Meanwhile, agency officials instructed the design team to follow two parallel paths – apply a preliminary system to help plan the FY92 budget and, for the longer term, develop a complete priority system with stakeholder participation. The requirement to conduct an application within a matter of weeks forced the design team to fall back on POS, which was applied without major modifications and, contrary to the ERG's request, without external participation. Concurrently, the design team wrote a 'conceptual design report' (USDOE, 1990) describing a modified version of POS, which was intended to serve as the straw man. The proposed system would be similar to the POS, but would include the uncertainty reduction and socio-economic criteria recommended by the ERG. The report identified opportunities for external involvement in the development and application of the proposed system, including eliciting value judgments from external groups as a mechanism for setting system weights and allowing external review or participation in the scoring process. However, no specific discussion was provided for how such participation would be obtained.

### 9.4.5   ERG response

The conceptual design report was distributed to the ERG in April, and the team conducted personal interviews to obtain ERG reactions. Not surprisingly, ERG members were angered by the decision to conduct an application without involving the ERG, and many objected to the design report's lack of a specific proposal for external involvement. As one member stated in an October 1990 letter written on behalf of the ERG,[5] the application of the preliminary system

would have been an excellent opportunity to meaningfully involve the ERG members ... consistent with the last principle cited in the ERG's February ... letter. No prioritization system will be acceptable if it does not provide for meaningful involvement by states, Tribes, and other interested parties in the implementation of the system...

Similar concerns were expressed at a third ERG meeting, held in September 1990. A plan was presented for revising the system to incorporate the features outlined in the conceptual design report, but ERG members were again more interested in discussing system use than in reviewing technical

---

[4] D.S. Miller, 'External Review Group comments on DOE prioritization system'. Letter to P. Whitfield dated February 9, 1990.
[5] D.S. Miller, letter to L.P. Duffy dated October 1990.

details. Several participants charged that the unstated intent of the system was to provide DOE with a justification for not seeking adequate funds to comply with triparty agreements. Despite the earlier STGWG poll indicating that simplicity was relatively unimportant, system complexity was raised as a major objection, and some members questioned whether a sophisticated system was justified given the poor quality of available risk data. One member argued that a system that the public could not understand could never provide a successful basis for public participation. Another speculated that DOE deliberately chose a complex methodology for the purpose of obfuscating real DOE goals. Finally, ERG members expressed indignation over the lack of a formal DOE response to their statement of February 9 and threatened to boycott future meetings unless a written response was provided.

### 9.4.6   ERPS implementation

Reacting to the negative ERG response, DOE promised that it would not use the system to develop the total budget request, that it would work to improve the quality of risk information in conjunction with the Agency for Toxic Substances and Disease Registry (a non-regulatory agency of the US Public Health Service), and that public participation in any future applications would be integrated with public participation in other EM programmes.[6]

The design team then attempted to modify the system further to accommodate ERG views. In response to principles nos 1 and 2 in the ERG's consensus statement, the three-tiered activity screening component was added, along with the constraint (which could be switched off to prevent infeasibility) that allowed the system to consider only funding allocations estimated by facility personnel to be sufficient to achieve full regulatory compliance. Simplifying the system and making the other changes desired by the ERG, however, proved problematic. An outside expert was asked to suggest ways to simplify the system. The expert concluded, 'While the system has enough components that it could be considered complex, it is hard to see what could be left out. All of the measures that are scored are important'.[7] Also, the design team did not view uncertainty and poor data quality as valid reasons for using unsophisticated methods. Formal methods such as value-of-information were selected precisely because they allow uncertainties to be taken into account and properly incorporated into the system logic.

Detailed plans for external involvement were developed, but a commitment from DOE to a specific approach could not be obtained. An unresolved issue was an Office of Management and Budget directive requiring DOE to protect as confidential certain pre-decisional information related to budgeting requirements (US Office of Management and Budget, 1990). The DOE

---

[6] Response to the February 9 'Statement of the External Review Group on the DOE Priority System', attachment to letters from L.P. Duffy to ERG members, dated December 9, 1990.
[7] C. Whipple, letter to P. Whitfield dated July 10, 1990.

reasoned that it was administratively restricted from allowing stakeholders access to information about activity costs prior to the establishment of the final budget by the President.

The system that was now ERPS was described in a revised design report (USDOE, 1991b). A discussion paper on public involvement was prepared, and a final meeting with the ERG was held in February 1991. At this meeting the design team provided a detailed presentation of ERPS. Citing the various changes that had been made in response to ERG comments, the DOE declared that the ERG had completed its task and no future meetings would be scheduled. Public involvement in future applications of ERPS was promised; however, responsibility and the specifics for public involvement were left to local field office and facility personnel. DOE also announced that a technical review of the system would be conducted by an independent group of technical experts and that a Federal Register Notice and workshop on the system were being planned. Although a few ERG members granted that the system had some merit, most insisted that DOE had ignored their concerns.

### 9.4.7   Additional applications and priority system developments

A reapplication of the system to FY92 planning was conducted during September–October 1990, and an application to FY93 planning was conducted during January–February 1991. Another application to FY93 planning was conducted during April–May 1991, and an application to support FY92 allocation decisions was conducted during July–August 1991. Meanwhile, in accordance with the plan to develop similar priority systems for other EM offices, the Office of Waste Management began developing a system, named the Resource Allocation Support System (RASS), for aiding the allocation of the waste management budget.

For the latter ERPS applications, at least some facilities involved external parties in the development or review of scores. However, DOE HQ played no role in this involvement and it was considered inadequate by many external parties. Application results showed that the inclusion of the uncertainty reduction and additional socio-economic criteria substantially altered system recommendations. Hanford, for example, was now being recommended for more, not less, funding. For the first time, DOE reported changing funding decisions in the directions recommended by the system. Specifically, for the FY92 application, DOE reported shifting $92 million of funding for 11 facilities and, for the FY93 application, about $80 million for seven facilities. According to the system, the funding changes made to the FY92 and FY93 allocations increased benefits by $300 million and $270 million, respectively.

### 9.4.8   Independent reviews and the Keystone Dialogue

The dissolution of the ERG did not put an end to active opposition. An ex-ERG member criticized the system in testimony delivered before Congress in the context of hearings on the FY92 budget (Werner and Reicher, 1991). He charged that ERPS was too complex and its use 'premature' because of inadequacies of site data, especially risk estimates. In some instances, critical comments were based on gross misunderstandings. For example, after receiving a half-day briefing, a review group reported that the 'proposed budget priority setting system is not satisfactory for its intended purpose'. This conclusion was based in part on the incorrect observation that 'the system does not provide a mechanism for determining the relative level of funding among the different sites. It only allocates funds within a site' (Advisory Committee on Nuclear Facility Safety to the Secretary of Energy, 1991). Another reviewer, under contract to an ERG member, provided a draft, 18 page critique that concluded, '[the system] shows a disregard for basic principles of economic rationality and utilitarian equity' (Nguyen, 1990). However, after the design team provided a personal briefing, the author retracted the statement and wrote, 'the additional information provided by DOE-HQ ... has theoretically supported the proposed framework of the priority system as a sound and fair mechanism to assist in the Environmental Restoration budget allocation process'.[8]

The Congressional Research Service (CRS) commented on the emerging priority system in a report dated February 6, 1991 (Holt *et al.*, 1991). It noted that many outsiders do not believe the risk estimates produced by DOE. The report also noted that the ERG had been critical of the system and that ERG members said that its recommendations had been largely ignored. Observing that the priority system was recommending substantial shifts in DOE clean-up funding, the report suggested that it was natural that ERPS would generate controversy.

### 9.4.8.1    The Technical Review Group

In April 1991, the promised independent technical peer review of ERPS was conducted. A Technical Review Group (TRG) was established, composed of 11 prominent experts in public health, hazardous waste management, decision analysis, hydrology, budget management and policy analysis. The TRG reviewed all available system documentation and received a 2 day briefing from the design team. Several ERG members were invited to the briefing, and they summarized the ERG's concerns.

The results of the TRG review were presented in a 28 page report (Burke *et al.*, 1991) with conclusions that, given the intensity of previous ERPS criticisms, came as a surprise to many. ERPS was deemed 'state of the art'. The report concluded:

This system represents an impressive intellectual accomplishment. Significant amounts of effective work have obviously been required to bring the process to the stage at which it could be employed in the field and produce results. Clever

[8] V.N. Nguyen, letter to T.P. Longo dated July 30, 1990.

innovations are found in many areas, and a long document could be written identifying admirable aspects of this system. The system is well-designed, technically competent, appropriate to its purpose, and ready for use...

The key to the system's design is its explicit acceptance of multiattribute utility as the best approach to such complex prioritization problems. The tool fits the problem very well. A formal, quantitative priority system is much to be preferred to an informal qualitative system. It allows replication of results, reduces subjectivity, and can help offset the lobbying and 'earmarking' that will inevitably occur in a political environment.

The TRG acknowledged that individuals without technical training might have difficulty understanding the basic workings of the system, but recommended developing simplified presentations and descriptions rather than simplifying the system itself.

Like the ERG, the TRG recommended that the system not be used as the basis for determining the size of the budget, but only to help allocate funds among facilities. The TRG also recognized the inherent conflict between ERPS and regulatory requirements:

The role of the tripartite agreements already signed was not clear. If abrogated, then the word of the government is worthless. If fully implemented, then the agreements can siphon all of the funds and a prioritization system is superfluous.

Despite recommending against using the system to set the total budget, the TRG added, 'one of the most valuable consequences of the use of the system would be a reassessment by DOE and the Congress of the overall costs and benefits of environmental restoration'.


### 9.4.8.2    Federal Register Notice, national workshops and additional reviews

Word that DOE was about to publicly unveil ERPS spread in early 1991. The General Accounting Office (GAO), a congressional oversight organization, told a congressional task force to expect DOE to publish a Federal Register Notice on ERPS within 'about 1 month'. Successfully implementing a workable system, the GAO said, was 'crucial' in view of 66 compliance agreements already signed and 22 more expected by the end of the fiscal year (GAO, 1991).

The promised Federal Register Notice finally appeared on September 8, 1991 (Federal Register Notice, 1991). To encourage comments, a 'national workshop' on the system was held in October. Discussion and comments submitted in response to the Federal Register Notice reflected earlier themes: system use, complexity, relationship to agreements and regulations, lack of a plan for public participation, and quality of inputs. Fourteen comment letters were received, ranging in length from 1 to 84 pages (USDOE, 1992). A small fraction were complimentary, but most repeated the earlier criticisms from the ERG. For example, ERPS was called 'far, far too complex' and, because of its 'callous disregard for the regulatory requirements of the law, ... probably illegal ...'.[9]

An ERPS application was scheduled for early 1992 to help plan the FY94 budget submission. A training session was held in January, but DOE terminated the application a couple of weeks later. The official explanation was that FY94 funds were estimated to be just adequate to achieve all applicable requirements and agreements, so there was no need to apply the system. It was obvious, however, that the system was in trouble. The GAO again reported on the system in congressional testimony delivered March 1992. The controversy surrounding the system was noted, and ERPS's future was called 'unclear' (Rezendes, 1992).

Meanwhile, a separate design team was proceeding with the development of RASS, the sister priority system for the Office of Waste Management. Hoping to avoid the problems encountered by ERPS, the team spent more than a year developing concept papers and conducting interviews and workshops with stakeholders, including some of the most vocal ERPS critics. Despite the incentives to make RASS look different, the technical advantages were such that the independent team chose a design mathematically identical to ERPS, with measures and scales appropriately tailored to the waste management context (Whitfield *et al.*, 1993). The system was successfully pilot tested in the summer of 1992 (Buehring *et al.*, 1992) and called technically 'sound' by an independent technical review group in December (Technical Review Group, 1992). However, external stakeholders raised the same objections to RASS that had been expressed about ERPS.

### 9.4.8.3   The National Academy of Sciences review

Faced with mounting criticism, DOE announced that it was requesting a detailed review of ERPS by the National Academy of Sciences (NAS). A major goal of the review was to resolve the conflicting opinions of the system. A special panel was established and a 30 month schedule adopted.

The first meeting of the panel was devoted to a detailed technical review of the system. Then, at the second meeting, the panel sought comments from facility representatives who had provided scoring inputs in past applications. Scorers said that they had found the exercise of providing inputs for ERPS useful for internal planning. When ERPS applications were terminated, their facilities developed their own internal priority systems to provide some of the benefits that had been lost.

### 9.4.8.4   The Keystone Report

The final events that led to the shelving of ERPS began with a report referred to as the Keystone Report, released in February 1993 by the Federal

---

[9] M.W. Grainey, letter to Leo Duffy dated November 5, 1991.

Facilities Environmental Restoration (FFER) Dialogue Committee (FFER, 1993). The committee, facilitated by the Keystone Center, was chartered by the EPA and tasked with developing consensus recommendations for improving the process by which environmental restoration decisions are made by federal agencies. Committee membership included representatives from federal agencies, including DOE and EPA; members of state governments; Native American representatives; and representatives from environmental, citizen and labour groups. Several ex-ERG members participated.

Obtaining consensus recommendations was reportedly extremely difficult due to distrust among the participating parties. However, after more than a year, the committee produced a draft report which included a recommendation for how to allocate ER funds among facilities in the event of funding shortfalls. The process, referred to as 'fair-share allocation', is highly detailed and emphasizes negotiation and consultation among stakeholders. Basically, however, the fair-share recommendation is that all agencies request budget levels sufficient to meet all their regulatory commitments, and that, if Congress fails to allocate those funds, the shortfalls would be shared proportionately among facilities: 'For example, if Congress only appropriates 90% of what was requested, then all sites ... would be expected to receive 90% of what was requested ...'. If conflict with legal obligations seems likely, then, according to the agreement, DOE may seek changes in the scope or schedule of clean-up activities. Regulators would renegotiate agreements and forego punitive enforcement actions as long as the department made good-faith efforts to follow the fair-share allocation process.

Obviously, ERPS was in conflict with the fair-share approach. As a participant in the Keystone process, DOE had, in effect, agreed to forego ERPS. The final ERPS analysis involved using data collected during the FY92 application to compare allocations recommended by the system with those that would be produced under fair share's proportional reduction. The results suggested that proportional reduction is considerably less effective at achieving ER objectives, with progressively greater losses the greater the funding shortfall. If proportional reduction had been used for FY92, the result, according to the system, would have been a lower average achievement according to every priority system criterion, including regulatory compliance, and roughly a $400 million loss in ER benefits.

Noting DOE participation in the development of the Keystone fair-share proposal, the NAS panel enquired whether there was any purpose in continuing its review of ERPS. On May 16, 1993, the DOE requested that the NAS terminate its review. Later, in testimony before Congress, Assistant Secretary Thomas Grumbly explained that the department had found that its formal priority systems were 'unworkable ... because of a combination of technical problems and lack of regulator and public involvement' (USDOE, 1994). 'The Department has not "abandoned" these systems', he clarified, 'but has deferred implementation pending further development' (Grumbly, 1993).

## 9.5   Lessons

Did ERPS fail? No, but ERPS certainly failed to achieve its potential. The pressures on DOE to use a national priority system continue to grow. Over the next 5 years, EM is anticipating a staggering $17 billion shortfall in funding needed to meet compliance agreements (Grumbly, 1995). DOE recently renounced fair-share allocation[10] and called on Congress to change the laws – reopen the compliance agreements and require that funding to sites be based on risks. Stakeholder resistance has not diminished. As Daryl Kimball of Physicians for Social Responsibility explained:[11]

What they are proposing is to dismantle a decision-making process that is based to a large extent on state and local input and involvement, and to replace it with a system of priority setting that is driven by Congress and the Department of Energy.

From a public involvement standpoint, ERPS was, of course, a failure. The ERPS experience provides a six-step plan for how *not* to do stakeholder involvement:

1. Pick a highly technical topic (risk-based priority setting) and a decision (budget allocation) not likely to be turned over to outsiders.

2. Propose external participation to develop something many external parties believe is unnecessary and undesirable (a way to allocate funds other than as needed to achieve compliance).

3. Pick external participants who have a personal stake against perceived system goals (e.g. people who helped negotiate agreements that the system appears designed to replace).

4. Begin the process by doing exactly the opposite of what the group demands (conduct an application without their involvement).

5. Keep doing it over and over again (additional applications).

6. Keep promising but don't deliver (public participation).

Whether ERPS erred in its balance between technical defensibility and simplicity is debatable. Consideration is now being given to using an alternative priority system designed to rank activities.[12] The system is simpler than ERPS, but is incapable of recommending funding allocations across sites. In addition to reduced capability, simplicity leads to other criticisms. A January 1995 letter to the Secretary of Energy,[13] signed by the attorneys general of 21 states and US territories, criticized DOE's renewed proposal for risk-based prioritization for being unresponsive to the myriad

---

[10] T.P. Grumbly, letter to G.A. Norton dated February 2, 1995.
[11] *New York Times*, December 21, 1994, p. A16.
[12] T.P. Grumbly, memo to Field Office Managers, Chief Financial Officers, and Assistant Managers for Environmental Management, dated January 31, 1995.
[13] G.A. Norton and 20 other attorneys general, letter to H.R. O'Leary dated January 6, 1995

site-specific factors that must also be considered in establishing priorities. Using the example of an activity that allows cost savings without reducing risks, the attorneys general argued that priorities are better set by individual facility managers and local regulators using 'best professional judgment'.

The failure of ERPS to gain support from states and local regulators is disappointing, given its unusual ability to account for site-specific factors and to incorporate local professional judgment. The system's many evaluation criteria allowed facility managers to account for virtually all relevant site-specific *information*, including opportunities for cost savings. Moreover, ERPS was responsive to site-specific *values*. Because ERPS evaluated and compared whole packages of activities, facility managers could preserve the priorities established with local regulators, something not possible with a centralized system for ranking activities. At the same time, ERPS provided a healthy incentive for work consistent with national values. Facility managers could propose funding cases inconsistent with national objectives, but they were discouraged from doing so because such cases compete less effectively in the ERPS competition for resources. These characteristics were never the subject of external comments.

One reason for the lack of comment on these presumably attractive features of ERPS was that they were never highlighted by DOE in describing or defending the system. Although an enormous amount of very detailed information was collected during each ERPS application, this information was not widely disseminated, either within DOE or to external stakeholders. Internally, the system suffered from confusion over its relationship with other DOE planning and priority setting tools. Although ERPS was designed specifically as a budgeting aid, it was implicitly linked with EM's 5 year planning process, which was not a budgeting exercise. Conflicts between DOE offices occurred over what systems should be used for prioritization. As discussed, ERPS was developed by the Office of Environmental Restoration (ER), but the Office of Environment, Safety, and Health has some oversight responsibilities for ER sites and uses an entirely different risk-ranking model to rank these same sites. Another problem was that midway through the project, the DOE project manager changed job responsibilities. As a result, ERPS lost a strong internal advocate with the technical and historical knowledge needed to most effectively promote and defend the system.

One clear lesson from ERPS is that favourable technical review does not necessarily lead to widespread acceptance. DOE was confident in the ability of the design team to develop a technically defensible system. POS, the predecessor system, was well regarded, and the senior members of the design team had previously received high marks from the NAS and other external technical reviewers in another prioritization effort based on a similar methodology (Merkhofer and Keeney, 1987). This confidence convinced DOE that ERPS was a relatively safe vehicle for external involvement and therefore a good means for responding to the pressures to open agency activities up to external review. Budget planning, however, was clearly not something that DOE was prepared to turn over to outsiders. Although DOE used the directive protecting pre-decisional information as an argument against external involvement in ERPS applications, the pilot test of RASS was conducted with outsiders. In this instance, DOE reasoned that it had complied with the order by expressing costs in approximate, rather than precise quantitative terms (Grumbly, 1995).

Unfavourable external opinion about ERPS was underscored by comments at a NAS workshop on DOE's clean-up programme. The workshop, part of a DOE-sponsored effort to improve the agency's use of risk assessment and risk management, was held to solicit views on how DOE had operated and should operate. During round-table discussion, an ex-ERG member expressed worry that DOE's interest in risk assessment and risk management might be a subterfuge for another agency attempt to impose ERPS, a system he described as receiving 'nearly universal criticism and condemnation'. The problem with the system, he clarified, was not so much the tool, but 'who used it and how it was used' (Miller, 1993).

This last statement provides, in the opinion of the authors, a concise summary of the main reasons for the controversy surrounding ERPS. External parties 'had no quarrel with the math', as some stated at the national workshop. Reviewers were prepared to agree that, if the system could be provided with accurate inputs, its outputs would be useful for decision making. What they questioned was whether the inputs to the system were accurate or could, in light of limited information, ever be of sufficient quality to justify the use of the system. More importantly, the basic function of the system (to allocate funds in a way other than in strict accordance with legal agreements) was not a desirable goal in the opinion of many external parties.

The problem was 'who used it'. That is, DOE used ERPS without external involvement. DOE charged the ERG with the job of helping to design the system but didn't listen well enough to what the ERG had to say. The ERG clearly recommended that DOE involve external parties and that DOE publish a public participation plan, but DOE did neither. Promises of public involvement were made, but DOE never approved a public participation plan for ERPS, even though draft plans were developed and pushed by members of the design team (JK Research Associates, 1992). Because a concrete plan for public involvement was promised but never delivered, stakeholders became increasingly frustrated by what they saw as DOE's unwillingness to interact meaningfully, while repeated applications were conducted without public involvement.

The problem was 'how it was used'. Many external parties were suspicious that the unstated purpose of the system was to justify DOE requesting a budget total that would be less than that needed to meet DOE's negotiated agreements. Outputs such as Fig. 9.5, with the designated one-to-one funding level, were interpreted as confirming their worst suspicions. DOE appeared poised to use the argument that spending what was required by its agreements was not cost-effective. Even when DOE promised not to use the system to set a total budget amount, external parties did not believe that such a policy would hold for long.

In summary, ERPS floundered not because of technical inadequacies, but because of the following:

- It became a lightning rod for the larger, still unresolved and still highly contentious issue of whether risk or compliance or some combination of the two should determine congressional funding for environmental restoration.

- It threatened the interests of powerful states with stringent compliance agreements. Such states were not anxious to explore whether their agreements might be skewing large ER budgets to their sites. Keystone's 'fair-share' approach was a less threatening way to handle the problem.

- DOE asked for the opinions of the ERG but did not respond to them. The ERG explicitly warned that an application of the system without public involvement would be unacceptable. DOE failed to involve interested parties in the first application of the system, thereby permanently alienating the ERG.

- DOE failed to provide adequate public involvement. Commitment to public involvement was expressed and verbal promises to publish a public participation plan were made, but there was never adequate follow-through.

An NAS committee charged with reviewing systems used by federal agencies to rank hazardous waste sites warned that 'conflicts' are likely between triparty compliance agreements and risk-based rankings. The committee then offered this comment (National Research Council, 1994b):

A system such as ERPS could be an important tool in helping to address such conflicts by providing a more objective evaluation of desired remedial priorities. Thus, while the committee is not in a position to judge whether the ERPS would provide the objective ranking of sites for remediation that is needed, it nevertheless believes that a well-developed, documented, validated and comprehensive model can help greatly in making sound decisions about what sites to remediate first, and the degree of cleanup that is desirable.

Hopefully, ERPS will provide an example for future priority system efforts. It suggests the potential that priority systems offer and illustrates some serious challenges that must be overcome.

# References

Advisory Committee on Nuclear Facility Safety to the Secretary of Energy 1991: Final report on DOE Nuclear Facilities, p. 98.

Applied Decision Analysis, Inc. 1993: *Instructions for generating inputs for the FY 1994 application of the DOE Environmental Restoration Priority System.*

Menlo Park, CA. (Earlier versions of this document supported earlier applications of the system.)

Buehring, W.A., Whitfield, R.G. and Wolsko, T.D. 1992: *Draft design report: a resource allocation support system for the U.S. Department of Energy*. Argonne, IL: Office of Waste Management, Argonne National Laboratory.

Burke, T., Dyer, J., Edwards, W., Hulebak, K., Matalas, N., Messner, H., Parker, F., Paulson, G., Russell, C., Russell, M. and Zeckhauser, R. 1991: *Report of the Technical Review Group of the Department of Energy's priority system for environmental restoration*. Washington, DC: DOE Environmental Restoration Prioritization System Independent Review Group.

Carnegie Commission on Science, Technology, and Government 1993: *Risk and the environment: improving regulatory decision making*. New York: Carnegie Corporation of New York.

Committee on Armed Services 1987: National Defense Authorization Act for Fiscal Year 1988–1989. *US House of Representatives, Report on H.R. 1748*, Public Law 100–180, April 15, 1987, 316–17.

Federal Facility Environmental Restoration Policy Dialogue Committee (FFER) 1993: *Interim report of the Federal Facility Environmental Restoration Dialogue Committee*. Keystone, CO: Keystone Center, Science and Public Policy Program.

Federal Register Notice 1991: Department of Energy request for public review and comment on a preliminary design report: a priority system for environmental restoration. Vol. **56**(173), 44 078–80.

GAO 1991: DOE to unveil system for setting defense waste cleanup priorities. *Inside Energy*, April 15, 6.

Grumbly, T.P. 1993: Statement of Thomas P. Grumbly, Assistant Secretary for Environmental Restoration and Waste Management, U.S. Department of Energy, before the Committee on Energy and Natural Resources, United States Senate, November 9, p. 5.

Grumbly, T.P. 1995: The Environmental Management Budget Challenge: working together to do more with less. Remarks delivered at the National Governors' Association Meeting, ANA Hotel, Washington, DC, February 7.

Holt, M., Meltz, R. and Simpson, M. 1991: Setting priorities for Department of Energy Environmental Activities. *CRS report for Congress*. Congressional Research Service.

Howard, R.A. 1966: Information value theory. *IEEE Transactions on Systems, Science and Cybernetics* **SSC-2**(1), 22–6.

Jenni, K., Merkhofer, M.W. and Williams, C. 1995: 'The rise and fall of a risk-based priority system: lessons from DOE's Environmental Restoration Priority System'. *Risk Analysis* **15**(3), 397–410.

JK Research Associates, Inc. 1992: *Environmental Restoration Priority System public participation plan*. Draft.

Keeney, R.W. and Raiffa, H. 1976: *Decisions with multiple objectives, preferences and value tradeoffs*. New York: Wiley.

Longo, T.P., Whitfield, R.P., Cotton, T.A. and Merkhofer, M.W. 1990: DOE's formal priority system for funding environmental cleanup. *Federal Facilities Environmental Journal* **1**(2), 221.

Merkhofer, M.W., Cotton, T.A., Jenni, K.E. Longo, T.P. and Lehr, J. 1989: A Program Optimization System for cleanup of DOE hazardous waste sites. *Proceedings of the Waste '89 Waste Management Symposium, Session XXVII*, March 2, Tucson, Arizona.

Merkhofer, M.W. and Keeney, R.L. 1987: A multiattribute utility analysis of alternative sites for the disposal of nuclear waste. *Risk Analysis* 7(2), 173–94.

Miller, D. 1993: Comment on videotape. *A workshop to review risk management in the Department of Energy's Environmental Remediation Program.* National Research Council, National Academy Press, Video Tape 3.

National Governors' Association and the National Association of Attorneys General 1990: *Report of the Task Force on Federal Facilities.* Washington, DC: National Governors' Association.

National Research Council 1994a: *Building consensus through risk assessment and management of the Department of Energy's Environmental Remediation Program.* Washington, DC: National Academy Press.

National Research Council 1994b: *Ranking hazardous waste sites for remedial action.* Committee on Remedial Action, Board on Environmental Studies and Toxicology, Commission on Geosciences, Environment, and Resources. Washington, DC: National Academy Press, 208–9.

Nguyen, V.N. 1990: Review of 'A preliminary conceptual design of a formal priority system for environmental restoration, Working Draft'. Minneapolis, MN: EWA, Inc.

Rezendes, V.S. 1992: Testimony before the Department of Energy Defense Nuclear Facilities Panel, Committee on Armed Services, House of Representatives. Energy Issues, Resources, Community, and Economic Development Division, General Accounting Office, GAO/T-RCED-92-43, p. 5.

Technical Review Group 1992: U.S. Department of Energy, Office of Waste Management Resource Allocation Support System (RASS) Technical Review Group meeting summary. Argonne, IL: Argonne National Laboratory.

US Department of Energy 1988a: *Environmental survey: preliminary summary report of the Defense Production Facilities.* DOE/EA-0072, Office of Environmental Audit, Environment, Safety and Health.

US Department of Energy 1988b: Priority System for Department of Energy Defense Complex Environmental Restoration Program. *A report prepared for the House Armed Service Committee.* Washington, DC: US Department of Energy, Defense Programs.

US Department of Energy 1989: *DOE News*: Watkins announces ten-point plan for environmental protection, waste management, R-89-068, p. 2.

US Department of Energy 1990: *A preliminary conceptual design of a formal priority system for environmental restoration.* Washington DC: Office of Environmental Restoration.

US Department of Energy 1991a: *Environmental Restoration Priority System fact sheet.* Revised Fall 1991. Washington, DC: USDOE.

US Department of Energy 1991b: *Preliminary design report: a formal priority system for environmental restoration.* Office of Environmental Restoration.

US Department of Energy 1992: *Priority System reading room information.* Office of Environmental Restoration and Waste Management, Office of Environmental Restoration.

US Department of Energy 1993: *Environmental restoration and waste management five-year plan, fiscal years 1994–1998: executive summary.* Washington, DC: USDOE.

US Department of Energy 1994: *FY 1995 Budget request overview and selected sites.* Briefing package for congressional staff and news media release. Office of Environmental Management. Washington, DC: USDOE.

US Environmental Protection Agency 1990a: *Reducing risk: setting priorities and strategies*. Washington, DC: USEPA.
US Environmental Protection Agency 1990b: *Hazard Ranking System, Final Rule*. 40 CFR Part 300 Federal Register, Vol. 55, 51 532–669.
US Environmental Protection Agency 1993: *A guidebook to comparing risks and setting environmental priorities*. Washington, DC: USEPA.
Voth, M.A., Merkhofer, M.W. and Jenni, K. 1993: *Priority System Technical Description*. ADA-93-2107. Applied Decision Analysis, Inc., Menlo Park, CA.
US Office of Management and Budget 1990: *Circular A-109*.
Werner, J.D. and Reicher, D.W. 1991: Statement of James D. Werner and Dan. W. Reicher, Esq., on behalf of the Natural Resources Defense Council before the Department of Energy Defense Nuclear Facilities Panel of the Committee on Armed Services US House of Representatives regarding the Department of Energy's Fiscal Year 1992 Budget, p. 42.
Whitfield, R.G., Buehring, W.A., Wolsko, T.D., Kier, P.H., Absil, M.J.G., Jusko, M.J. and Sapinski, P.F. 1993: *Resource allocation support system (RASS): summary report of the 1992 pilot study*. Argonne, IL: Argonne National Laboratory.

# 10 Model elicitation in competitive markets

C M Queen

## 10.1  Introduction

In a competitive market there are several brands of the same product which compete together for custom. The competition in the market can cause complex relationships to exist between the various brand sales in the market. If we are to find a realistic model of the brand sales, we need a multivariate model which models these relationships explicitly. The nature of the problem also specifies several other properties which a practical model would ideally have. In this paper we shall use the structure of the problem of modelling in competitive markets to elicit a suitable statistical model for the brand sales. Firstly let us consider the structure of the problem.

Even in markets which are highly competitive, not all brands in the market necessarily compete with all the other brands in the market. Instead, what often happens is that the brands form themselves into competing *submarkets* in which the brands within the same submarket compete together more heavily than they do with the rest of the market as a whole (Day *et al.*, 1979). For example, in the washing powder market we may have a submarket made up of all 'environmental' washing powders and another submarket made up of all the 'colour care' washing powders. So, for example, an 'environmental' washing powder will compete more heavily with 'environmental' powders than with other powders in the market. Now, these submarkets can be overlapping so that at least one brand competes in more than one submarket (Arabie *et al.*, 1981; Srivastava *et al.*, 1984). For example, all those brands which are both 'environmental' and 'colour care' will compete in both submarkets, causing these submarkets to overlap. By identifying the competing overlapping submarkets in a market, we can get some understanding of how the brands are related to one another and therefore what type of multivariate structure we need our elicited statistical model to have.

In addition to requiring a model which is able to accommodate complex brand relationships, there are other features which companies would ideally like any proposed model to have. The first feature which our model should ideally have is parsimony. Although we need a highly multivariate model, it needs to be relatively simple to implement if these models are to be of use to companies in practice. The second feature required is interpretability. Any parameters in the model need to be easily interpretable to enable market researchers to input information and their expert opinion of the market situation into the model. Finally, the models should ideally be adaptable to changing market situations both in terms of sales changing rapidly through time and in terms of brands leaving and new brands entering the market place.

This paper will illustrate how we can use this information about the problem and the type of model we need to elicit a suitable statistical model. An obvious starting place is to investigate how the brand relationships in an overlapping market manifest themselves through the brand sales.

Suppose that we have an overlapping market with $m$ brands and we want to model the brand sales $Y_1, ..., Y_m$. Following on from Queen *et al.* (1994) and Queen (1994), we can think of $\mathbf{Y}$ as a multinomial sample so that

$$p(\mathbf{y} \mid \boldsymbol{\psi}, N) = \frac{N!}{\displaystyle\prod_{j=1}^{m} y_j!} \prod_{j=1}^{m} \psi_j^{y_j} \tag{10.1}$$

where

$$N = \sum_{j=1}^{m} y_j$$

$$\psi_j = P(\text{customer chooses brand } j), \qquad j = 1, ..., m$$

So if $N \sim Po(\mu)$, then $Y_j \sim Po(\mu \psi_j)$, such that $\sum_{j=1}^{m} \psi_j = 1$. We could then model $\mathbf{Y}$ by modelling $\psi$ and $\mu$. Finding a suitable prior for $\mu$ is straightforward (see Queen *et al.*, 1994). However, finding a suitable prior for $\psi$ which accommodates an overlapping market structure is not so easy.

Suppose that the brands are classified into $n$ submarkets labelled $C(1), ..., C(n)$. Suppose further that we classify consumers into $n$ different types, labelled $T(1), ..., T(n)$, so that a type $T(i)$ consumer will only choose between brands in submarket $C(i)$. Thus we are classifying consumers according to their purchase behaviour. Then the probability that brand $j$ is chosen by a consumer will depend on what customer type the consumer is, and so

$$\psi_j = \sum_{i=1}^{n} P[\text{customer chooses brand } j \mid \text{customer is type } T(i)] \times$$

$$P[\text{customer is type } T(i)]$$

So, if we have an overlapping market, when modelling ψ we really want to model the brand choice probabilities for each brand and customer and $P$[customer is type $T(i)$], for each type. However, we can only observe $m$ sales at once, whereas for this model we would need to estimate more than $m$ probabilities. Therefore, as the model stands, it is overparametrized.

It will be shown how we can use the overlapping structure of a market to reparametrize ψ so that brand relationships imposed by the overlapping market are modelled explicitly, but so that the problem is not overparametrized. The methods will be illustrated by considering a particular case study of eliciting a model for the breakfast cereal market. There are many possible ways to elicit the overlapping submarkets in a market. In Section 10.2 we will consider one such possible method using data of customer purchases in the breakfast cereal market. These submarkets are then used in Section 10.3 to find a suitable reparametrization of ψ which respects the brand relationships in the market but does not overparametrize the problem. This reparametrization is then used to elicit a Bayesian model of the market.

## 10.2 Eliciting the market structure for the breakfast cereal market

One way of thinking of the breakfast cereal market is to divide it up into three groups – adult/family brands, children's brands or bran-based brands. Hammond *et al.* (1996) give details of a study of consumer behaviour of the three leading brands in each of these breakfast cereal groups. This study consisted of panel data of customers in Britain in 1988 in which a researcher went into each panellist's home and did an inventory of the breakfast cereals the panellist had bought. The percentage of customers of each brand $B_j$ who also bought brand $B_k$ was then recorded for $j \neq k$. The results are displayed in Table 10.1.

The information in Table 10.1 can be used to elicit the competing submarkets in this market. We will categorize consumers into various types, $T(1), \ldots, T(n)$, such that all customers of the same type choose between the same brands. Thus we will categorize customers according to which group of brands they choose between. The brands between which type $T(i)$ chooses then form competing submarket $C(i)$.

From Table 10.1, we could say that type $T(i)$ chooses between both brands $B_j$ and $B_k$ if the percentage of $B_j$ customers who also buy $B_k$ is large in comparison to the general percentage level of other customers who also buy $B_k$, and vice versa. This essentially means that brand $B_j$ customers are more likely also to buy brand $B_k$ than most of the other customers in general, and similarly $B_k$ customers are more likely also to buy $B_j$ than most of the other customers in general. We are therefore trying to identify pairs of brands, $B_j$ and $B_k$, between which there is a lot of brand switching. If there is a lot of brand switching, this would suggest that there is a type of customer who switches between $B_j$ and $B_k$ so that they compete together in a submarket.

**Table 10.1** Annual duplications of purchase for the 3 leading brands in each brand grouping in the ready-to-eat cereals market. A value with an asterisk means that this value is particularly large in comparison to the other values in that column. (Taken from Hammond *et al.*, 1996.)

| Customers of: | Percentage who also bought: | | | | | | | | |
|---|---|---|---|---|---|---|---|---|---|
| | Adult/family | | | Children's | | | Bran-based | | |
| | Corn Flakes | Shredded Wheat | Alpen | Rice Krispies | Shreddies | Coco Pops | K. Bran Flakes | All Bran | PL Bran Flakes |
| *Adult/family* | | | | | | | | | |
| $B_1$ = K. Corn Flakes | – | 27 | 15 | 43 | 23 | 11 | 17 | 16 | 13 |
| $B_2$ = Shredded Wheat | 61 | – | 16 | 46 | 37* | 11 | 18 | 17 | 17 |
| $B_3$ = Alpen | 59 | 29 | – | 37 | 25 | 14 | 21 | 20 | 14 |
| *Children's* | | | | | | | | | |
| $B_4$ = Rice Krispies | 68 | 32 | 14 | – | 34* | 17* | 18 | 15 | 14 |
| $B_5$ = Shreddies | 63 | 44* | 17 | 58* | – | 19* | 20 | 15 | 18 |
| $B_6$ = K. Coco Pops | 66 | 28 | 20 | 61* | 40* | – | 19 | 14 | 12 |
| *Bran-based* | | | | | | | | | |
| $B_7$ = K. Bran Flakes | 58 | 26 | 17 | 37 | 24 | 11 | – | 28* | 30* |
| $B_8$ = K. All Bran | 54 | 26 | 17 | 33 | 19 | 8 | 30* | – | 22* |
| $B_9$ = PL Bran Flakes | 48 | 27 | 13 | 34 | 26 | 8 | 34* | 23* | – |

In terms of Table 10.1, we can easily identify which brands $B_j$ and $B_k$ have the above property by identifying any percentages which are large in relation to the other percentages in their column. All of these relatively large percentages are marked with an asterisk in Table 10.1. For example it can be seen that Shreddies customers are much more likely also to buy Shredded Wheat than the other customers are, and so the value of 44 has an asterisk. Notice also that the percentage of Shredded Wheat customers who also buy Shreddies is also large in comparison with the majority of the other customer percentages. This therefore suggests that there is a lot of brand switching between Shreddies and Shredded Wheat, which in turn indicates that Shreddies and Shredded Wheat are perhaps competing in the same submarket and are purchased by the same type of purchaser. So by using the asterisk values in Table 10.1, we can define three types of purchaser:

$T(1)$ = children's purchaser $\Rightarrow T(1)$ chooses between $\{B_4, B_5, B_6\}$

$T(2)$ = bran-based purchaser $\Rightarrow T(2)$ chooses between $\{B_7, B_8, B_9\}$

$T(3)$ = 'shredd' purchaser $\quad \Rightarrow T(3)$ chooses between $\{B_2, B_5\}$

We still have two brands, $B_1$ and $B_3$, which have not been identified as being bought by any of these three types of customer. Now, customers of Alpen $(B_3)$ don't buy any other brand more than any other customers and no other customers particularly favour buying Alpen. We could therefore define a further customer type:

$T(4)$ = Alpen purchaser $\Rightarrow T(4)$ chooses $B_3$

Notice that a high percentage of *all* customers also buy K. Corn Flakes, which might suggest that each customer type chooses $B_1$ *in addition* to the other brands identified above. This suggests that we could finally define the competing submarkets as:

$$C(1) = \{B_1, B_4, B_5, B_6\}$$
$$C(2) = \{B_1, B_7, B_8, B_9\}$$
$$C(3) = \{B_1, B_2, B_5\}$$
$$C(4) = \{B_1, B_3\}$$

The competition in a market can be represented pictorially by a market graph (Queen *et al.*, 1994). In a market graph each brand is represented by a node on the graph and two nodes, $B_j$ and $B_k$, are joined together by an undirected edge if $B_j$ and $B_k$ compete together in a submarket. A market graph gives a simple and clear picture of the market. The market graph for the breakfast cereal market is given in Fig. 10.1.

## 10.3 Eliciting a model for a competitive market

Once we have elicited a market structure for a competitive market, we can use this information to build a model of the sales in the market. Recall from

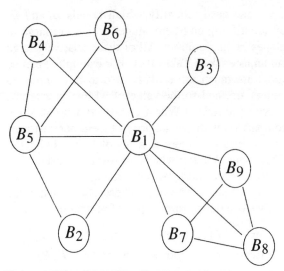

**Fig. 10.1** Market graph for the breakfast cereal market

Section 10.1 that if we want to model $\psi$ realistically in a competitive market, we need to reparametrize $\psi$ so that

- the reparametrization respects the overlapping structure of the market;
- we don't overparametrize the problem.

The market graph can help us identify such a parametrization. To illustrate how this may be done, let us consider the breakfast cereal market again.

### 10.3.1   Reparametrizing choice probabilities in the breakfast cereal market

If we are to model the breakfast cereal market realistically, we need to reparametrize $\psi_1, \ldots, \psi_9$ by using the information about the competing submarkets elicited in Section 10.2. The market graph can help us in visualizing such a reparametrization. Notice that $B_1$ competes with all other brands in the market. So to start our reparametrization of $\psi$, define

$$P(\text{consumer chooses } B_1) = 1 - \lambda$$

Consider submarket $C(1)$. Now

$$P(\text{consumer chooses } B_4, B_5 \text{ or } B_6)$$
$$= P(\text{consumer chooses } B_4, B_5 \text{ or } B_6 \mid \text{not chosen } B_1)P(\text{not chosen } B_1)$$
$$+ P(\text{consumer chooses } B_4, B_5 \text{ or } B_6 \mid \text{chosen } B_1)P(\text{chosen } B_1)$$

So if we let

$$P(\text{consumer chooses } B_4, B_5 \text{ or } B_6 \mid \text{not chosen } B_1) = \theta(1)$$

then

$$P(\text{consumer chooses } B_4, B_5 \text{ or } B_6) = \theta(1)\lambda$$

Similarly, for the other submarkets, let

$$P(\text{consumer chooses } B_7, B_8 \text{ or } B_9 \,|\, \text{not chosen } B_1) = \theta(2)$$
$$P(\text{consumer chooses } B_2 \text{ or } B_5 \,|\, \text{not chosen } B_1) = \theta(3)$$
$$P(\text{consumer chooses } B_3 \,|\, \text{not chosen } B_1) = \theta(4)$$

Then

$$P(\text{consumer chooses } B_7, B_8, B_9) = \theta(2)\lambda$$
$$P(\text{consumer chooses } B_2, B_5) = \theta(3)\lambda$$
$$P(\text{consumer chooses } B_3) = \theta(4)\lambda$$

(see Fig. 10.2).

We have therefore reparametrized $\psi_1$ and $\psi_3$ in terms of our new parameters $\lambda$ and $\theta$, so that

$$\psi_1 = P(\text{consumer chooses } B_1) = 1 - \lambda$$
$$\psi_3 = P(\text{consumer chooses } B_3) = \theta(4)\lambda$$

To reparametrize the rest of $\psi$, we will use the brand groupings defined by the submarkets. $\psi_7$, $\psi_8$ and $\psi_9$ are the easiest to reparametrize out of the remaining brands, so we will tackle these first. Now, $P(\text{consumer chooses}$

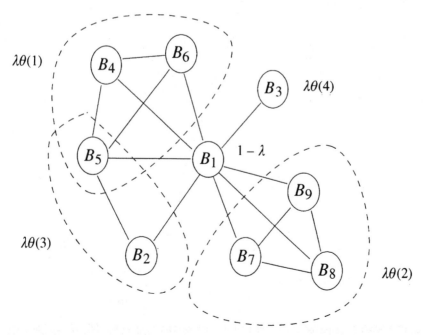

**Fig. 10.2** Market graph labelling the probabilities of choosing groups of brands

$B_7$) can be expressed as

$P$(consumer chooses $B_7$)

$\quad = P$(consumer chooses $B_7\,|$ consumer chooses $B_7, B_8, B_9$)

$\quad\quad \times P$(consumer chooses $B_7, B_8, B_9$)

and $\psi_8$ and $\psi_9$ can be expressed similarly. Therefore if we define

$P$(consumer chooses $B_7\,|$ consumer chooses $B_7, B_8, B_9$) $= \rho_2(1)$

$P$(consumer chooses $B_8\,|$ consumer chooses $B_7, B_8, B_9$) $= \rho_2(2)$

$P$(consumer chooses $B_9\,|$ consumer chooses $B_7, B_8, B_9$) $=$

$$1 - \rho_2(1) - \rho_2(2)$$

where $0 < \rho_2(1), \rho_2(2) < 1$ (see Fig. 10.3), then

$\psi_7 = \rho_2(1)\theta(2)\lambda$

$\psi_8 = \rho_2(2)\theta(2)\lambda$

$\psi_9 = [1 - \rho_2(1) - \rho_2(2)]\theta(2)\lambda$

(Note that the subscript 2 for $\rho_2$ is to match $\theta(2)$.)

Finding a suitable reparametrization of the remaining parameters is a little more difficult. The main problem is that $B_5$ is bought by both $T(1)$ and $T(3)$.

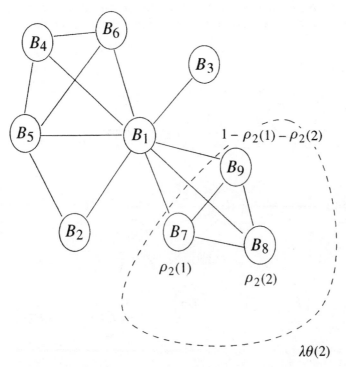

**Fig. 10.3** Market graph labelling the probabilities of choosing group $\{B_7, B_8, B_9\}$ and the conditional probabilities *within* this group

Ideally we would use exactly the same idea as we did for $B_7$, $B_8$ and $B_9$ and define

$P(\text{consumer chooses } B_4 \mid \text{consumer chooses } B_4, B_5, B_6) = \rho_1(1)$

$P(\text{consumer chooses } B_5 \mid \text{consumer chooses } B_4, B_5, B_6) = \rho_1(2)$

$P(\text{consumer chooses } B_6 \mid \text{consumer chooses } B_4, B_5, B_6) = 1 - \rho_1(1) - \rho_1(2)$

$P(\text{consumer chooses } B_2 \mid \text{consumer chooses } B_2, B_5) = \rho_3$

$P(\text{consumer chooses } B_5 \mid \text{consumer chooses } B_2, B_5) = 1 - \rho_3$

where all the $\rho$ parameters are $>0$ and $<1$. Then $\psi_4$, $\psi_6$ and $\psi_2$ have a similar form to $\psi_7$, $\psi_8$ and $\psi_9$ above, although this time

$$\psi_5 = [\rho_1(2)\theta(1) + (1 - \rho_3)\theta(3)]\lambda$$

(see Fig. 10.4). However, the problem with this reparametrization is that we now have too many parameters. There are nine parameters $(\lambda, \theta(1), \theta(2), \theta(3), \rho_1(1), \rho_1(2), \rho_2(1), \rho_2(2), \rho_3)$ which we wish to estimate. However, we only have nine time series, and since we are

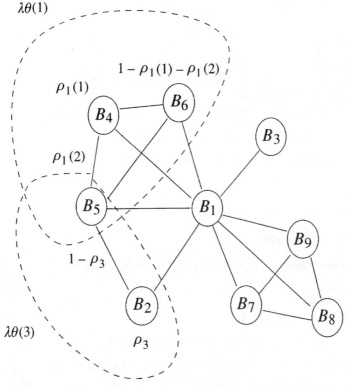

**Fig. 10.4** Market graph labelling the probabilities of choosing groups $\{B_2, B_5\}$ or $\{B_4, B_5, B_6\}$ and the conditional probabilities *within* these groups

conditioning on the total sales in the multinomial likelihood, this means that we only have enough data to learn about eight parameters. Therefore if we are to find an identifiable model, we need to reduce the number of unknown parameters.

One way around this overparametrization problem is to put constraints on parameters which have little available information about them, so that the number of unknown parameters can be reduced. The *differentially non-informedness hypothesis* (d.n.h.) (Rubin, 1976; Dawid and Dickey, 1977) sets all parameters with little information about them equal, unless there is sufficient evidence from the data to dispute this. In our example, the parameters which will often have little information about them will be $\rho_1(2)$ and $1 - \rho_3$, as it can be difficult to learn about the probability of buying a brand for different customer types. Therefore using the d.n.h., we will assume here that

$P$(consumer chooses $B_5$ | consumer chooses $B_4, B_5, B_6$)

$\quad = P$(consumer chooses $B_5$ | consumer chooses $B_2, B_5$)

i.e.

$P$(consumer chooses $B_5$ | type $T(1)$ chooses)

$\quad = P$(consumer chooses $B_5$ | type $T(3)$ chooses)

Using the d.n.h. is, of course, not ideal. However, its use *does* allow us to elicit a model which *almost* accommodates the overlapping structure, and yet is not overparametrized. Note that if good information about $\rho_1(2)$ and $1 - \rho_3$ does exist, we could use this so that we wouldn't have to assume the d.n.h. (see Queen, 1991, for details).

We can now use the d.n.h. to reparametrize $\psi$. Firstly, as above let

$P$(consumer chooses $B_2$ | consumer chooses $B_2, B_5$) $= \rho_3$

$P$(consumer chooses $B_5$ | consumer chooses $B_2, B_5$) $= 1 - \rho_3$

But now set

$P$(consumer chooses $B_5$ | consumer chooses $B_4, B_5, B_6$) $= 1 - \rho_3$

also. Because of this constraint,

$P$(consumer chooses $B_4$ or $B_6$ | consumer chooses $B_4, B_5, B_6$) $= \rho_3$

(see Fig. 10.5). We can then write

$\psi_4 = P$(consumer chooses $B_4$)

$\quad = P$(consumer chooses $B_4$ | consumer chooses $B_4, B_6$)

$\quad \times P$(consumer chooses $B_4$ or $B_6$ | consumer chooses $B_4, B_5, B_6$)

$\quad \times P$(consumer chooses $B_4, B_5, B_6$)

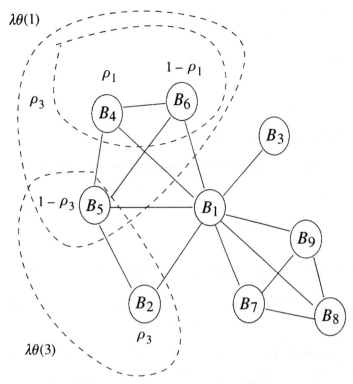

**Fig. 10.5** Market graph labelling the probabilities of choosing groups $\{B_2, B_5\}$ or $\{B_4, B_5, B_6\}$ and the conditional probabilities within these groups using the d.n.h.

and $\psi_6$ can be written similarly. Therefore defining

$$P(\text{consumer chooses } B_4 \,|\, \text{consumer chooses } B_4, B_6) = \rho_1$$
$$P(\text{consumer chooses } B_6 \,|\, \text{consumer chooses } B_4, B_6) = 1 - \rho_1$$

where $0 < \rho_1 < 1$, then

$$\psi_2 = \rho_3 \theta(3) \lambda$$
$$\psi_4 = \rho_1 \rho_3 \theta(1) \lambda$$
$$\psi_5 = (1 - \rho_3)[\theta(3) + \theta(1)] \lambda$$
$$\psi_6 = (1 - \rho_1) \rho_3 \theta(1) \lambda$$

We have thus managed to reparametrize the eight-dimensional parameter $\psi$ in terms of the eight new parameters $(\lambda, \theta(1), \theta(2), \theta(3), \rho_1, \rho_2(1), \rho_2(2), \rho_3)$. As this new parametrization was found using the competitive groupings of brands in the market, it provides a realistic model of the choice probabilities in the market.

### 10.3.2 A Bayesian model of the market

Now that we have reparametrized $\psi$, we can re-express the likelihood given in eq. (10.1) in terms of these new parameters, so that:

$$l(\psi\,|\,y) \propto \prod_{i=1}^{9} \psi_j^{y_j}$$

$$\equiv (1-\lambda)^{y_1}(\rho_3\theta(3)\lambda)^{y_2}(\theta(4)\lambda)^{y_3}(\rho_1\rho_3\theta(1)\lambda)^{y_4}((1-\rho_3)(\theta(3)+\theta(1))\lambda)^{y_5}$$

$$\times ((1-\rho_1)\rho_3\theta(1)\lambda)^{y_6}(\rho_2(1)\theta(2)\lambda)^{y_7}(\rho_2(2)\theta(2)\lambda)^{y_8}((1-\rho_2(1)$$

$$-\rho_2(2))\theta(2)\lambda)^{y_9}$$

$$= [(1-\lambda)^{y_1}\lambda^{\sum_{j=2}^{9}y_j}][\rho_1^{y_4}(1-\rho_1)^{y_6}][\rho_2(1)^{y_7}\rho_2(2)^{y_8}(1-\rho_2(1)$$

$$-\rho_2(2))^{y_9}][\rho_3^{y_2+y_4+y_6}(1-\rho_3)^{y_5}]$$

$$\times [\theta(1)^{y_4+y_6}\theta(3)^{y_2}(\theta(3)+\theta(1))^{y_5}\theta(2)^{y_7+y_8+y_9}\theta(4)^{y_3}]$$

This is essentially the product of separate likelihoods for $\lambda$, $\rho_1$, $\rho_2$, $\rho_3$ and $\theta$. Now, in a Bayesian analysis, as we collect more and more data, the posterior will become more like the likelihood. But, if this is the case, then the posterior will indicate that each of these parameters is in fact independent. So, as the parameters will become independent as we observe data, for ease of computation it is sensible to assume that these parameters start independent. Thus we can set *independent* priors for each set of parameters.

Clearly $l(\lambda\,|\,y)$, $l(\rho_1\,|\,y)$ and $l(\rho_3\,|\,y)$ are binomial likelihoods and so beta priors would be used as conjugate priors for these parameters. $l(\rho_2\,|\,y)$ is a multinomial likelihood and so the natural conjugate prior to use is the Dirichlet of the form:

$$p(\rho_2) \propto \rho_2(1)^{a_1-1}\rho_2(2)^{a_2-1}[1-\rho_2(1)-\rho_2(2)]^{a_3-1} \qquad a_j>0, j=1,2,3$$

The posterior for $\rho_2$ will then be Dirichlet with updated parameters $a_1+y_7$, $a_2+y_8$ and $a_3+y_9$. The natural conjugate prior to use for $\theta$ has the form:

$$p(\theta) \propto \theta(1)^{\gamma_1}\theta(3)^{\gamma_2}(\theta(3)+\theta(1))^{\gamma_3}\theta(2)^{\gamma_4}\theta(4)^{\gamma_5}$$

for $\gamma_j>0$, $j=1,\ldots,4$. This is called the generalized Dirichlet prior. The posterior for $\theta$ will again be generalized Dirichlet with updated parameters $\gamma_1+y_4+y_6$, $\gamma_2+y_2$, $\gamma_3+y_5$, $\gamma_4+y_7+y_8+y_9$ and $\gamma_5+y_3$.

We have therefore been able to use the market structure, as expressed by the overlapping submarkets, to define a Bayesian model which can accommodate the brand relationships in the market. By defining independent conjugate priors for each separate set of parameters, we now have a model which not only accommodates a complex multivariate structure, but which is also simple enough to be implemented in practice.

The parameters in the model have a natural interpretation. For example, $\theta(1)$ is the proportion of sales in subgroup $\{B_4, B_5, B_6\}$ within $C(i)$, $\rho_2$ are the market shares within the competing subgroup of brands $B_7, B_8, B_9$, and $\rho_1$ is the market share of $B_4$ within the group $B_4, B_6$. Notice that the grouping of $B_4$ and $B_6$ imposed by the d.n.h. actually seems a natural subgrouping within $C(1)$ because of $B_5$'s competition in $C(3)$. By using the interpretation of the model parameters, it is possible for companies to utilize their expert knowledge of markets effectively. Queen *et al.* (1994) demonstrate exactly how expert knowledge can be incorporated into the model.

As mentioned in Section 10.1, any model of brand sales which is to be useful to companies must ideally be able to adapt to changing market situations. In terms of modelling sales levels which are changing over time, following the ideas presented in Queen (1994) and Queen *et al.* (1994), it can be shown how the Bayesian model proposed here can easily form the basis of a multivariate Bayesian dynamic time series model. Sudden dramatic changes in sales levels can be easily modelled by using subjective intervention within the time series framework. Both Queen (1994) and Queen *et al.* (1994) demonstrate how this can be done. Preliminary use of these time series models for modelling two other competitive markets has been extremely promising (Queen, 1994; Queen *et al.*, 1994).

Adapting the model to changes in the market due to brands leaving or entering the market is a little more difficult. For every different market graph we will have a separate reparametrization of $\psi$ and so we will define a different model. Despite this, however, if the market changes we might be able to salvage some parts of our old model. Let us illustrate how this might be done by considering an example.

Suppose that we have a new brand, $B_{10}$, entering the market which we believe competes in submarket $C(4)$. The market graph for this new market is given in Fig. 10.6. Using the same kind of reparametrization ideas introduced earlier, to accommodate this new brand we would need to define a new parameter

$$P(\text{consumer chooses } B_3 \mid \text{consumer chooses } B_3, B_{10}) = \rho_4$$
$$P(\text{consumer chooses } B_{10} \mid \text{consumer chooses } B_3, B_{10}) = 1 - \rho_4$$

This new parameter would then be incorporated into the model in the same way as all the others. However, the parameters in the rest of the model will be unaffected by this change and we would not have to discard all the information which we had so far learnt about these parameters.

Another way in which the market graph could alter is if a brand changed its competitive position within the market. For example, suppose that $B_3$ started to compete in the 'bran' market and moved into submarket $C(2)$. This time we now assume that there are only three types of consumer so that $T(2)$ and $T(4)$ have merged. We could then carry forward information about the old $\theta(2)$ and $\theta(4)$ when learning about the new merged type probability. Once again, the parameters associated with $C(1)$ and $C(3)$ would remain unaffected by this change in the market. The market graph is given in Fig. 10.7.

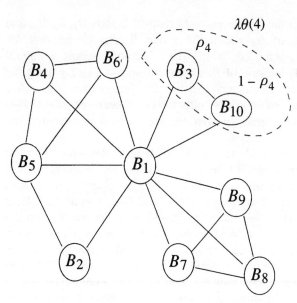

**Fig. 10.6** Market graph with new brand $B_{10}$ entering submarket $C(4)$ and with corresponding extra parameters

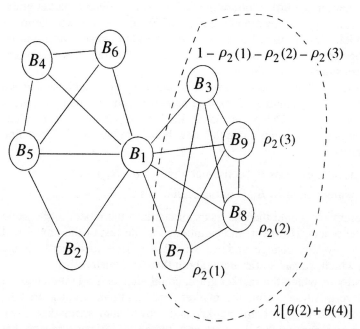

**Fig. 10.7** Market graph when $B_3$ starts to compete in $C(2)$ with corresponding new parameters

## 10.4 Conclusion

In this paper we have used the given structure of the problem of modelling in competitive markets to elicit a suitable statistical model. By using the problem structure when specifying the model, we have managed to define a model which exhibits most of the features which we ideally wanted from the model. We have shown here how such a model can be elicited for the breakfast cereal market. The use of these models in other markets has so far been very promising.

# References

Arabie, P., Carroll, J.D., DeSarbo, W. and Wind, Y. 1981: Overlapping clustering: a new method for product positioning. *Journal of Marketing Research*, **18**, 310–17.

Dawid, A.P. and Dickey, J.M. 1977: Likelihood and Bayesian inference from selectively reported data. *Journal of the American Statistical Association* **72**, 845–50.

Day, G., Shocker, A. and Srivastava, R. 1979: Consumer oriented approaches to identifying product markets. *Journal of Marketing* **43**, 8–19.

Hammond, K., Ehrenberg, A.S.C. and Goodhardt, G.J. 1996: Market segmentation for competitive brand. *European Journal of Marketing* **30**, 39–49.

Queen, C.M. 1991: Bayesian graphical forecasting models for business time series. *PhD thesis*, University of Warwick.

Queen, C.M. 1994: Using the multiregression dynamic model to forecast the brand sales in a competitive market. *The Statistician* **43**(1), 87–98.

Queen, C.M., Smith, J.Q. and James, D.M. 1994: Bayesian forecasts in markets with overlapping structures. *International Journal of Forecasting* **10**, 209–33.

Rubin, D.B. 1976: Inference and missing data. *Biometrika* **63**, 581–92.

Srivastava, R.K., Alpert, M.I. and Shocker, A.D. 1984: A customer-oriented approach for determining market structures. *Journal of Marketing* **48**, 32–45.

# 11   Building a Bayesian model in a scientific environment: managing uncertainty after an accident[1]

D Ranyard, J Q Smith

## 11.1   Introduction

In 1990 we were invited to a meeting to discuss how to reconcile physical measurements with the predictions of a complicated deterministic model in the early stages of an accidental release of radioactive substances after an accident at a nuclear power plant. The work would be supported by the European Union who planned to develop software which would support emergency response in many different countries in both western and eastern europe.

The demands of the project were fierce. Firstly, they required an automatic mechanism, supported by theoretical justification, which would adapt forecasts of the geographical spread of contamination in the light of various measurements that might be taken. Secondly, they wanted realistic confidence bounds on these adjusted forecasts. A third requirement was that the uncertainty management was fast: they demanded that the computational time was of the same order of magnitude as the running of the forecast model without uncertainty management. A fourth criterion was that any methodology was adaptable to the particular needs of each country involved in the project, in particular to the type and extent of monitoring data that might arrive. A fifth demand was that the method was consistent and clamped on to the physicists' deterministic models already partially devised. Finally, any method needed to be reasonably transparent to the scientific user.

In this chapter, we will outline some of the more successful methods we used to first elicit a Bayesian model for the structure of this problem and then to build a methodology which provided computationally feasible forecasts of

[1] The views expressed in this paper are those of the authors and do not necessarily reflect those of the RODOS project.

required quantities. Obviously, for reasons of space we will limit our discussion to a subset of the problems we addressed. In any elicitation exercise, information tends to arrive in a rather jumbled fashion, but again, for clarity, we have set out the procedure in its most logical order. Although the elicitation methodology described here is by no means meant as a template, we hope that it will be helpful to readers facing a similar exercise.

In any project of this sort it is first necessary to obtain at least a reasonable understanding of the science underlying the physical models being used. Fortunately some close collaborators had been working for several years in this area so at least we knew who to ask to obtain different kinds of information. In Section 11.2 we will give you a little of this background information. Next it is important to elicit what is generally agreed about the scientific nature of the process, in this case the dispersal of toxic substances. There were many different dispersal models, all of which needed uncertainty management, but all had certain features in common. It was imperative that the uncertainty management respected these shared features, not only to make it compatible with classes of deterministic models but also to ensure that the introduction of uncertainty made scientific sense. This structure is outlined in Section 11.3.

Next it was necessary to have a precise idea of how it was envisaged the system might be used. We discovered our remit was to produce software to advise on whether certain conurbations should be evacuated, advised to shelter, or issued with iodine tablets in the first 24 hours of a release. So it became clear that the primary focus must lie in estimating the probabilities of dangerously high contamination both in space and time. Fortunately the term 'dangerously high' had been defined precisely in other components of the system and so was relatively specific.

The next part of the process of elicitation is to determine the sources of uncertainty. Scientists have a good understanding of some of these sources, particularly those associated with the physics. Engineers, who work; in the physical conditions plants, have a good understanding of the different ones; in particular they understand better what kind of information will arrive, when and how. Both the elicitation of the specification of the system and the sources of uncertainty are discussed in Section 11.4.

At this point, and only at this point, is it timely to begin to build a statistical model. Not only will it need to address all the issues mentioned above, but it will also need to be computationally feasible. How this was done and the compromises that needed to be made are outlined in Section 11.5.

Our project is not completed and various technical problems still need to be resolved. However we will illustrate how the system currently works and how it will be improved in Section 11.6.

We shall begin then, with an outline of the scientific setting.

## 11.2  Dispersal models

There are several different methods of modelling atmospheric dispersion, each with its own characteristics and suited to particular situations.

However, all models are based on a similar set of parameters:

- *Source term data* – this is in the form of estimates of source parameters such as height and strength of the release. It may be provided by experts, monitors inside the stack, or a mixture of both.

- *Dispersal patterns* – equations have been designed which model how a particular substance disperses in the atmosphere.

- *Meteorological data* – this information is averaged over time and must be interpolated between meteorological stations. The following information is provided:

  - the wind field, updated every 2 hours;
  - forecast of precipitation (and possibly satellite photographs);
  - stability class, an indication of how stable or unstable the weather conditions are.

- *Geographic information* – most atmospheric dispersion models require basic terrain data. Some have been developed further to work with complex terrain.

Models are only an approximation to real world processes and it is important to realize that they may diverge significantly from reality. Our intention in this project was to develop ways of handling the uncertainty associated with this divergence and thus keep the models on track.

The list above ignores the fact that in an emergency other information, most notably monitoring data, will also be available, giving still more information about the release. Monitoring data are not rigidly defined. They may not arrive in chronological order and may contain errors. Depending on the number of monitoring stations in the locality, these data may be sparse both spatially and temporally. They are recorded by permanent and mobile measurement stations which together provide:

- instantaneous air and ground readings

- air and ground readings, integrated over time

- gamma dose rates.

It may not be advisable to rely solely on these data, just as it would not be sensible to rely solely on the predictions of a model. We believed that by combining both methods we might get a clearer picture of what is actually happening and still have the ability to predict future events based on the current situation.

Atmospheric dispersion models fall into two broad categories: Lagrangian and Eulerian. Lagrangian models are usually numeric and simulate the trajectories of many emitted particles which follow a stochastic differential equation whose parameters are uncertain and need to be estimated. Experience suggests that even simple versions of these models cannot provide accurate real time predictions quickly enough for our purposes.

Eulerian models use a diffusion/advection equation to predict dispersion. By imposing rather severe restrictions on these models it is possible to obtain an analytic steady-state model which could be used as the basis for parameter estimation. The Gaussian plume model (Päsler-Sauer, 1986) assumes stationarity, a constant wind vector and homogeneous terrain. These restrictions make the model too simplistic to provide useful results in the early stages of an accident. However, a solution to this problem, posed by several authors including Mikkelsen *et al.* (1984), is to use a puff model.

Instead of assuming a continuous release from the source, it is assumed that the mass is released in a series of discrete puffs. These puffs can then be transported and dispersed around the local terrain based on the current wind field and local terrain (*see* Fig. 11.1) This method has been incorporated into the Risø-Meso-scale Puff model, RIMPUFF (Thykier-Nielsen and Mikkelsen, 1991).

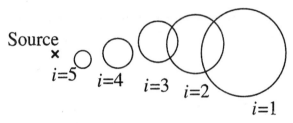

**Fig. 11.1** A puff model

To add to the accuracy of the RIMPUFF model, its designers added a further level of detail, puff splitting or *pentafurcation*. As puffs are released and transported over the local terrain, they grow in size (*see* Fig. 11.2).

When their diameter reaches a chosen threshold, they can split into five smaller puffs. The mass associated with the parent puff is distributed amongst the children which are also smaller in size. The relationship between this mass distribution and air concentration is modelled by, for example:

$$r_t(\mathbf{s}) = \frac{1}{(2\pi)^{2/3}\,\sigma_t(1)\,\sigma_t(2)\,\sigma_t(3)}$$
$$\exp\left\{-1/2\left[\frac{(s_1 - u_t(1))^2}{\sigma_t(1)^2} + \frac{(s_2 - u_t(2))^2}{\sigma_t(2)^2} + \frac{(s_3 - h)^2}{\sigma_t(3)^2}\right]\right\} \quad (11.1)$$

Here $(u_t(1), u_t(2))$, is a wind velocity vector at time $t$, $\mathbf{h}$ is the hetght of the emission and $\sigma(1)$, $\sigma(2)$ and $\sigma(3)$ govern the widening of the puff with time. So the contribution a particular fragment makes to a given site $\underline{s}$ at time $t$ is represented by a complicated equation $r$. The definition of $r$ may vary from country to country depending on the scientists involved, so we have had to be flexible and build our statistical model independently of it. A more detailed description may be found in Smith and French (1993).

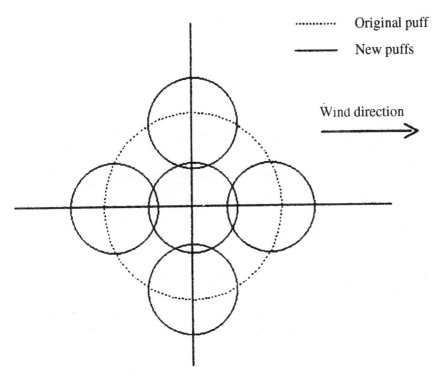

**Fig. 11.2** Puff pentafurcation

## 11.3 The agreed features of predictive dispersal models

From the last section it is clear that all dispersal models are specified conditional on it being possible to measure certain meteorological information. At the time of the accident, such information will usually be interpolated from a coarse grid. They will also use a variety of hard-wired topological information such as the position of various hills and valleys.

Secondly, all models are based on the assumption that they are conditionally Markov. By this we mean that were the dispersion profile of the contaminated gas to be known precisely at time $t$ then the subsequent position of the profile of that mass would not depend upon its dispersion profile at earlier times Of course this Markov structure is conditional on the inputs mentioned in the last paragraph. Thirdly, all dispersal models that are used assume that emissions occur from a single source and, once released, the mass of radioactive material is conserved.

Finally, it is generally accepted that most casualties of the first 12–24 hours will be caused by the inhalation of material contaminated with beta emissions. This radioactive contamination is then quickly transferred to vulnerable parts of the body, particularly the thyroid gland. Air concentration readings relate most closely to these beta emissions, although other readings such as the gamma dose rate and gamma spectrum give very useful indirect information about it.

It was imperative that any statistical techniques respected these basic assumptions.

## 11.4    Sources of uncertainty: scientists versus engineers

From the elicitation process, it became clear that uncertainty was associated with four distinct sources. First there was the uncertainty associated with the actual emergency event before any data were taken account of. Thus when an accident happens it is often clear that one of a small set of scenarios has occurred, each with an associated uncertain consequence relating to the spread of the plume. The physics of such scenarios is currently being coded and can be used as the basis of a prior distribution of what is happening at the source of an accident. Such prior information represents expert judgements from engineers and, to a lesser extent, physicists.

Once the accident has occurred, measurements begin to be taken. The precise nature of these measurements often depends on the plant and the conditions surrounding the accident. For example, gamma dose, gamma spectrum readings and instantaneous air concentration readings might be made available from the chimney stack and around the perimeter of the power plant. The gamma doses and instantaneous air concentration readings are obtained fairly quickly, whereas the gamma spectrum takes more time to analyse. Information of these types is also often available near vulnerable conurbations close to the site. In addition, some of the information collected depends on the country and type of installation, but, for example, in many countries vans will track the plume of contamination and take readings at sites where contamination might be expected to be high.

Measurement errors can occur in one of three ways. There are errors associated with instruments when they are used in appropriate circumstances and in the proper way. For example, air concentration is an aggregate measure which might be expected to exhibit Poisson (or a more overdispersed distribution's) variation. This type of variation is understood quite well by the physicist. In some situations a second source of error occurs because instruments give ambiguous readings. For example, gamma dose rate measurements can be grossly distorted because of physical problems like reflection. There may not be enough information to determine whether such distortions are occurring. Engineers tend to understand best this type of error. Finally, instruments may be misused. In an accident like Chernobyl it was observed that many of the people volunteering for the dangerous task of taking measurements were unclear how to use the instrumentation properly. Such errors are clearly associated with personnel and apparently can be expected to occur relatively frequently in an accident scenario.

A third source of model uncertainty concerns the atmospheric conditions. Although dispersal models assume conditions are given at all times and positions, wind field measurements, for example, are typically interpolated from a large, irregular grid, and therefore only a coarse approximation of their true values can be supplied. Local atmospheric features add to these errors. For example, rainfall can often be patchy, and whether there has been rain or not has a big effect on the amount of contamination still in the air. On top of this, demographic conditions are often uncertain. Many installations are by the sea and so, particularly in summer, the weekend population distribution is quite different from the weekday population. This information will rarely be recorded, however, although the efficacy of, for example, evacuation of the area will be very dependent on it.

The final, and perhaps the most disturbing, fact is that there are many different physical models of dispersion. The same physical experiments have been performed where the source emissions are controlled and known and meteorological conditions are known far better than they would be in actual accident. Despite this, results show that all these physical models give far from perfect results and are all different from one another. The current situation is that any statistical model which assumed a particular dispersal model was 'right' would in practice not work very well. There is a tendency by scientists to see all of uncertainty management in terms of the estimation of the parameters within their models which are themselves imperfect. This attitude encourages overoptimism until data deviate from the expected structure under the model when there is nothing left to work with. Engineers, on the other hand, being conscious of the practical difficulties that might occur in an emergency, tend to want to disregard most of the sophisticated physical models, despite the fact that these give a good insight into what might be happening and also provide good forecasts in stable conditions.

## 11.5 Building a statistical model

Early in the elicitation process it became clear that the structure of the problem was inherently high-dimensional and non-linear. However, the natural solution of using stochastic numerical techniques such as MCMC (e.g. Smith and Roberts, 1993; George *et al.*, 1994) would not be appropriate, not only because probability predictions need to be continually updated very quickly but also because of the regular prior information which needed to be coded in the system on-line. The nature of uncertainty in the structure of the model suggests it would be desirable if a statistical model had two features:

1. the model should adjust its predictions appropriately to accommodate incoming data;
2. various possible scenarios could be incorporated and the statistical model automatically adapted as information supported some hypotheses whilst contradicting others.

The natural way to address feature (1) would be to allow the statistical model to have at least two levels. Within the first stage observational error would be modelled, and in the second initially unpredicted stochastic change in the actual structure of the model would be accommodated. This would therefore require a hierarchical structure to the statistical model (see, e.g., O'Hagan, 1994, Section 9.5). Feature (2) was most simple to introduce by using a Bayesian mixture model.

The second imperative was that the structural hypothesis, particularly the spatial Markov property, mass conservation and the single source hypothesis were respected. The last hypothesis was actually a very helpful feature since it demanded that the temporal-spatial process could be seen as the artefact of a single time series, a model of emissions of contaminated mass from the chimney source. Bayesian models of hierarchical time series, especially the DLM, are well understood (see, for example, West and Harrison, 1989). In addition, we could expect that prior information, some quantitative and some qualitative, would be available from the time of the initial release.

At the time we started the analysis, only two physical models ran quickly enough to be incorporated into the project, and both these models were puff models. Because puff models release masses at discrete times, it was natural to model the emission process as a discrete time series. Chimney stack air concentration readings, $Y_t$ at time $t$, would be a measurement of the air concentration at that time, in the chimney, together with some observational error, so we have

$$Y_t = Q_t + \varepsilon_t(0) \tag{11.2}$$

where $\{Q_t : t = 1, 2, 3 \dots\}$ label the masses under each puff of mass concentration and $\{\varepsilon_t(0) : t = 1, 2, 3 \dots\}$ label independent observational errors.

Now, clearly the mass of concentration under adjacent puffs is likely to be highly positively correlated. With no information as to how the source will develop, an obvious candidate to model this dependence is to assume $\{Q_t : t = 1, 2, 3 \dots\}$ form a random walk

$$Q_t = Q_{t-1} + \eta_t \tag{11.3}$$

where $\{\eta_t : t = 2, 3, \dots\}$ are independent Gaussian errors with zero mean and variance $Z$. In fact, expert judgement often suggests that air concentration readings will increase exponentially at a known rate to an asymptote, at least approximately where the asymptote depends on the severity of the accident. This can be modelled by introducing another state parameter into eq. (11.3). Thus we set

$$Q_t = Q_{t-1} + R_t + \eta_t$$
$$R_t = \lambda R_{t-1} \tag{11.4}$$

where $0 \leqslant \lambda \leqslant 1$ and $R_t$ is the increase in air concentration between each puff emission. The mean and variance of $Q_1$ and $R_1$ and the variance $Z$ can be fixed to expert judgement about the expected asymptotic value of the emission and the uncertainty associated with it. As information from the stack arrives, so this asymptotic value is adjusted and variance reduced.

Details of how such matching of expert judgement and expected emission profile takes place are discussed in detail in Gargoum (1997, 1996).

The simplest data stream to initially model were air concentration readings taken at various times $t$ from the chimney (site $\underline{s} = \underline{0}$), see eq. (11.2) and other sites $\underline{s}$ away from the release. At other sites we can just use the aggregation formula eq. (11.1), again introducing an error, i.e.

$$Y_t(s) = \sum_{i=1}^{n(t)} r_t(s, i)Q(i) + \varepsilon_t(\underline{s}) \tag{11.5}$$

where $\{Q(i)\}$ are mass fragments which are uncertain, $r_t(s, i)$, $1 \leqslant i \leqslant n(t)$ are assumed known, and $\varepsilon_t(\underline{s})$ are error terms. There are some technical issues here: for example, $\varepsilon_t(0)$ and $\varepsilon_t(\underline{s})$ probably should not be Gaussian or should at least need to follow some variance law related to the mean of $Y_t(s)$. These are beyond the scope of this article but are discussed in more detail in Smith and French (1993) and Gargoum and Smith (1997). Suffice to say that Poisson-like variation, overdispersion and heterogeneity in spatial distribution can all be accommodated in a fairly straightforward way. Similar formulae relate gamma dose rate readings to masses.

The next issue to address was the uncertainty associated with the conditions surrounding the release. Now, because of extensive experimentation with deterministic atmospheric models, we were fortunate that the sensitivity of their forecasts to parameter misspecification was well understood. Typically, such models were most sensitive to the height of the release $h$ and the wind velocity $\underline{v}$. A natural and simple solution for representing this uncertainty was to place a discrete distribution over a grid of values on $(h, \underline{v})$, using expert judgements to assign prior probabilities $\Pi(h, \underline{v})$ over the set of values $(h, \underline{v})$ on the grid. The dispersal model $M(h, \underline{v})$ was run for each vector of parameter values $(h, \underline{v})$ for the set of observations $y$ that had been taken. Bayes' rule then allowed us to update the probabilities $\Pi(h, \underline{v})$ to $\Pi(h, \underline{v} \mid y)$ using the formula $\Pi(h, \underline{v})$ and

$$\Pi(h, \underline{v} \mid y) \propto \Pi(h, \underline{v})p(y \mid M(h, \underline{v})) \tag{11.6}$$

where the proportionality constant would be fixed to ensure that the posterior probabilities summed to 1. As new information arrived we could update again using $\Pi(h, \underline{v} \mid y)$ instead of $\Pi(h, \underline{v})$ and $p(y' \mid M(h, \underline{v}))$. The true contamination $\theta_t(\underline{s}, \bar{h}, \underline{v})$, assuming a height parameter $h$ and velocity $\underline{v}$, would be given by

$$\theta_t(\underline{s}, h, \underline{v}) = \sum_{i=1}^{n(t)} r_t(s, i, h, \underline{v})Q(i, h, \underline{v}) \tag{11.7}$$

It follows that the predictive density $p_t(\theta \mid \underline{s})$ of contamination at site $\underline{s}$, time $t$, would be given by the simple mixing formula

$$p_t(\theta \mid \underline{s}) = \sum_{(h, \underline{v})} \Pi(h, \underline{v} \mid y)p_t(\theta \mid \underline{s}, h, \underline{v}) \tag{11.8}$$

where $p_t(\theta \,|\, \underline{s}, h, \underline{v})$ is the density of $\theta_t(\underline{s}, h, \underline{v})$ given $\underline{y}$. This can be calculated by using Bayes' rule on the mass fragments $Q(i, h, \underline{v})$, $i = 1, 2, ..., n(t)$, conditioning on a set of independent observations of the four given in eq. (11.5). In particular, if the posterior mean of $Q_t(\underline{s}, h, \underline{v})$ was $\mu_t(\underline{s}, h, \underline{v})$, then the expected contamination at site $\underline{s}$, time $t$, would be given by the linear formula

$$\mu_t(\underline{s}) = \sum_{(h, v)} \Pi(h, \underline{v} \,|\, \underline{y}) \mu_t(\underline{s}, h, \underline{v}) \qquad (11.9)$$

For further discussion of mixture distributions in Bayesian modelling see Draper (1995).

Mixture modelling of this type is very useful but has two drawbacks. The first is that, if the prior distributions over the model space are at all complicated, they can be quite difficult to elicit reliably. The second is that they are computationally time-consuming since separate dispersion models need to be run for each parameter configuration. It was found that, to capture local turbulence, it was possible to make a probabilistic version of the pentafurcation process. Thus eq. (11.3) is replaced by the probabilistic version:

$$\underline{Q}_i^f = \rho Q_i + \underline{\omega}_i(t)$$

where $\underline{Q}_i^f$ are the vector of five fragments of an original puff of mass $Q_i$ and $\underline{\omega}_i(t)$ is a vector of errors such that $\underline{\omega}_i(t)(1, 1, ..., 1) = 0$, i.e. such that mass in the fragments totals the mass in the original puffs. Suitable covariance matrices for $\omega_i(t)$ are given in Smith and French (1993). The larger the setting of the variances in the covariance matrix $\Omega$, the greater the turbulence relative to the sampling error, and hence the greater the uncertainty as to how the puff would distribute its mass over its five fragments.

Finally, we needed to address the problem of the inherent uncertainty associated with the family of dispersal models being used. Of course some of this uncertainty is implicitly being modelled by the mixing given above, together with the error attributed to the pentafurcation process. Uncertainty about the relative weights we should give to data, model forecasts, the systematic nature of the release and local turbulence was modelled by mixing over various combinations of variances in the model. Different model predictions could be compared simply by including those different dispersal models in the mixture described above, and forecast variances could be systematically updated using a slight modification of the usual Bayesian conjugate analysis (see Gargoum, 1995). But even with all these features, it was important to systematically monitor the performance of the statistical model. In particular, if the data were not in accord with the predictions of the model it was necessary to know this quickly. The central checks of this were obviously the monitored data. Within any model we use a diagnostic which compares observations $y_t(\underline{s})$ with their predicted values $\hat{y}_t(\underline{s})$. We systematically plot $y_t(\underline{s}) - \hat{y}_t(\underline{s})$, the predictive errors of the model, and also $\sigma_t^{-1}(\underline{s})(y_t(\underline{s}) - \hat{y}_t(\underline{s}))$, the predicted errors normalized by the appropriate standard deviation. Large errors are plotted over a spatial grid so that any systematic discrepancy between the predicted and actual dispersal can be identified.

This simple statistical model has several appealing features. First, it is consistent with the scientific structural hypotheses outlined earlier. For example, given *precise* spatial information at time *t* of the distribution of the mass of contamination, all other information about the previous history of that mass would be independent of the model forecasts. Secondly, provided the mixing distributions are kept simple, to work adequately the model requires just a few extra parameters to be specified: variances which determine the relative weight that should be placed on errors associated with various uncertainties in model specification and model errors associated with unreliable data. These parameters are fairly easy to elicit (or at least their distribution is), because they relate directly to judgements the experts make anyway. Thirdly, of great comfort to the physicist is the knowledge that the statistical model retains the forecasts of the deterministic model as a special case: we just assign infinite variance to observational errors. So it is easy for the engineer and physicist to appreciate the compromises the statistical model makes to the data. The structure of the statistical model allows it to adapt to information *if and when* it arrives. If observations are missing, its forecast errors just increase; if data arrive late, this can be accommodated using Bayes' rule which is an algorithm independent of the order in which data is input. Finally we have, at least to some extent, modelled all sources of uncertainty in the process and can demonstrate this to be so to the sceptical engineer.

However, having built a theoretical framework around which uncertainty can be managed, we were confronted with inevitable computational problems. We had a mixture of many statistical models and with each of these components was associated a high-dimensional Gaussian process on a vector of puff fragment masses. The dimension of this vector increased quickly with time (exponentially) in a rather complicated way that depended on the dispersal model used and the atmospheric conditions that were observed. Furthermore, these masses were typically highly correlated with one another. How would it be possible to continually and quickly update the forecasts of the statistical model in real time in the light of sequences of new information? This is the subject of the next section.

## 11.6 Implementing the statistical model

### 11.6.1 Building the model

Initially, we built our model into the RIMPUFF code using a traditional Kalman filter to represent the correlation between puffs, source term and observed data. The only data assimilated were air concentration data. However, this presented computational problems as the number of puffs and observations increased. Basically, this method was too slow – we needed to provide model updating in real-time, similar to running the model without updating.

We turned to Bayesian belief nets to overcome this. By utilizing some of the known dependencies within the model we were able to represent the distribution as a dynamically changing belief net, developing and changing as puffs are released and pentafurcate and as observations are incorporated. This improved the efficiency of the system by an order of magnitude and met our target of updating the model in the same time magnitude as the model without updating. Some dynamic approximations had to be developed to incorporate realistic observations, but these proved to be reasonable (Gargoum and Smith, 1997; Smith and Papamichail, 1996; Gargoum, 1996).

### 11.6.2   Validation: model and method

Validation has not proved to be an easy task for the simple reason that very few real data sets exist. It is not difficult to synthesize observations from our model and then show that it will converge to these data given different initial parameters. This goes some way to proving a system, but what we would like to do is to use real data from tracer experiments. There are currently some problems associated with this:

1. very few suitable experiments exist;
2. those that do exist are typically very simple; such experiments have only ever been carried out on days with remarkably simple weather conditions;
3. it is not always obvious how to interpret observation data from these experiments, which usually come in the form of hour-long time-integrated measurements.

Air concentration is not the only data source. In particular, gamma dose rates are often measured close to the source. Golubentov *et al.* (1996) discuss appropriate algebraic relationships between air concentration measures and gamma dose, and Duranova *et al.* (1996) have developed ways of reconstructing source term activity from gamma data radial monitors using bootstrap techniques.

### 11.6.3   Further work

The relationship between another type of observation, the gamma spectrum and source activity has also been studied by Golubentov *et al.* (1996). It looks as if it will be relatively straightforward now, therefore, to accommodate data other than air concentration readings into our methodology and to produce multivariate predictions about specific radiation levels (e.g. iodine, noble gases, caesium, etc.) to enhance the decision aiding facilities of the system.

Much of the data collected are essentially count data. We are therefore in the process of investigating the use of other (non-normal) distributions, in particular Poisson distributions to describe the updating process. This seems to enhance forecasting ability, especially in areas of as yet low concentration.

Obviously, this prediction model is not intended to be used in isolation, and in the near future we intend to hook it up to a source term model which will provide estimates of the source term along with associated uncertainties. Further, the output of our model will become the input for a number of countermeasure and consequence models which will help with off-site emergency management of nuclear accidents. This forms part of the vision of the RODOS project, more details of which can be found in Kelly and Ehrhardt (1995).

Finally, we intend to look beyond short-range dispersion modelling and attempt to develop data assimilation methods for long-range forecasts. We must also look further ahead in time. Months and years after an accident, predictions will be based solely on observation data and we need to develop ways of moving smoothly from the prediction based models described here to such long-term models built over static data.

## References

French, S., Ranyard, D.C. and Smith, J.Q. 1995: *Uncertainty in RODOS*. School of Computer Studies, University of Leeds Research Report.

Gargoum, A.S. and Smith, J.Q. 1994: *Approximating dynamic junction forests*. Research report, Department of Statistics, University of Warwick.

Gargoum, A.S. and Smith, J.Q. 1997: *Dynamic generalised linear junction trees*. (Under revision for *Biometrika*.)

Gargoum, A.S. 1997: Bayesian dynamic models for the emission profile toxic gases. (Submitted to the *Journal of Forecasting*.)

Gargoum, A.S. 1996: Issues in Bayesian forecasting of dispersal after a nuclear accident. *PhD thesis*, University of Warwick.

George, E.I., Makov, U.E. and Smith, A.F.M. 1994: Fully Bayesian hierarchical analysis. In Freeman, P.R. and Smith, A.F.M. (eds) *Aspects of uncertainty*. Wiley, 191–8.

Golubentov, A., Borodin, R., Sohier, A. and Palma, C. 1996: Data assimilation and source term estimation during the early phase of a nuclear accident. *Report BLG 707 SCK.CEN.*

Kelly, G. N. and Ehrhardt, J. 1995: RODOS – A comprehensive decision support system for off-site emergency management. *Paper presented at the Fifth Topical Meeting on Emergency Preparedness and Response*. American Nuclear Society, Savannah, Georgia.

Mikkelsen, T., Larsen, S.E. and Thykier-Neilsen, S. 1984: Description of the Risø puff diffusion model. *Nuclear Safety* 67, 56–65.

O'Hagan, A. 1994: *Bayesian inference*. London: Edward Arnold.

Päsler-Sauer, J. 1986: Comparative calculations and validation studies with atmospheric dispersion models. *KfK Research Report*, 4146 Kernforschungzentrum, Karlsruhe.

Smith, J.Q. and French, S. 1993: Bayesian updating of atmospheric dispersion models for use after an accidental release of radiation. *The Statistician* 42, 501–11.

Smith, J.Q., French, S. and Ranyard, D.C. 1995: An efficient graphical algorithm for updating estimates of the dispersal of gaseous waste after an accidental release. In Gammerman A. (ed.) *Probabilistic reasoning and Bayesian belief networks*. Henley-on-Thames: Alfred Waller, 125–44.

Smith, J.Q. and Papamichail, K.N. 1996: *Fast Bayes and the dynamic junction forest*. Submitted to Artificial Intelligence.

Smith, A.F.M. and Roberts, G.O. 1993: Bayesian computations via the Gibbs sampler and related Markov chain Monte Carlo methods (with discussion). *Journal of the Royal Statistical Society* **B55**, 3–24.

Thykier-Nielsen, S. and Mikkelsen, T. 1991: *Rimpuff User Guide: Version 30*, Risø National Laboratory, Denmark.

West, M. and Harrison, P.J. 1989: *Bayesian forecasting and dynamic models*. New York: Springer Verlag.

# 12 Project evaluation in drug development: a case study in collaboration[1]

P J Regan, S Senn

## 12.1 Introduction

Pharmaceutical drug development, with its long time horizons, large investments, and significant scientific and commercial uncertainties, has long been an area of professional decision analysis practice, with some companies developing internal capabilities and others relying, as necessary, on outside professionals. Rising drug development costs, more competitive global drug markets and greater health care cost pressure have combined in recent years to highlight the value of a systematic approach to drug development.

This chapter describes the collaboration between a global pharmaceutical company, Ciba-Geigy (Ciba), which made the commitment to develop its own decision-analytic capability, and a professional decision consulting firm, Strategic Decisions Group (SDG), which adapted its existing approach to Ciba's unique needs. The collaboration is yielding a more systematic view of the drug development portfolio that facilitates communication between project teams and decision boards about success criteria, key scientific hurdles, investment requirements and commercial potential.

[1] We thank members of the action team that initiated the project evaluation effort at Ciba: Anders Hove, William Huebner, Ron Steele and Keith Widdowson. We also thank the steering committee members – William Jenkins, Martin AbEgg, and Elaine Snowhill – who supported this effort, and the project evaluation core team members – Terry Cook, Stefan Schmitt and Linda Shepard from Ciba; Robin Arnold and George Corrigan from SDG – who implemented it. Finally, we thank development project team members and others within Planning and Program Management and SDG who assisted in this effort.

The chapter covers:

- the perspectives of each organization at the outset of the collaboration;

- the motivation for a decision-analytic approach to project evaluation;

- the model templates devised to provide consistency and transparency;

- the primary benefit of improved communication rather than analytical sophistication;

- the process challenges involved in establishing decision analysis as an insightful communication language within the organization.

Section 12.2 describes the evolution of Ciba's perceived need for improved project evaluation capability as an outgrowth of a broader initiative. Section 12.3 describes the relationship of Ciba's situation to the broader pharmaceutical R&D decision analysis challenge, introducing a framework for the analytical and organizational challenges described in subsequent sections. Section 12.4 describes the productivity index as a measure of project value and introduces two examples and an analogy to sharpen insight into the portfolio management challenge.

Section 12.5 describes the model templates designed to bridge the gap between the high level information needed to determine the productivity index for a project and the narrower areas of expertise of individual members of the drug development project teams. Section 12.6 describes the project evaluation process for an illustrative example in step-wise fashion, focusing on the 'softer' aspects of the joint work, such as preparation, the sequence of meetings with the development project team, sharing of facilitation responsibilities by the client core team and the consultants, review by the decision board and follow-up.

Sections 12.7 and 12.8 summarize lessons learned and future challenges, respectively.

## 12.2   Background from the pharmaceutical company's perspective

Ciba underwent a thorough review of its drug development process which instigated a major programme for change under the heading 'Faster time to market' (FTTM). In many instances, FTTM was tactical in its focus, for example identifying ways in which the processes of planning, conducting and reporting clinical trials could be streamlined. Amongst the many subprojects carried out under the FTTM aegis, however, there are several with a more strategic emphasis. One of these, to develop a management tool to assist with project evaluation and prioritization, as well as the assessment of licensing-in opportunities, provides the background to this paper. The objective was to prepare for each project a project evaluation document (PED) assessing its worth. Accordingly, the team working on this methodology soon became known as 'the PED team'.

As might be expected, although FTTM had identified project evaluation as an important area for improvement, this did not mean that the problem was being ignored in the company. Existing methods had rightly concentrated on evaluating projects on the basis of important factors affecting their value, in particular expected marketed return if successful and probability of success. However, the existing approaches had a number of weaknesses, as follows:

- little explicit attention was paid to development costs;

- the methods tended to use weighting schemes which, however extensively based on preliminary investigations to establish their reasonableness, lacked a firm basis in logic;

- logical inconsistency prevented company-wide consensus on what should be done. This in its turn meant that local variants of the general approach were in use.

These approaches, which were arguably adequate in the past, could not be allowed to continue. Changing conditions in the international pharmaceutical market, as well as rapidly increasing development costs (in part a result of ever more demanding regulatory requirements), meant that costs would have to become a much more important part of any prioritization system. Furthermore, global development was now an imperative. Drug candidates could no longer be developed semi-independently in Europe and the United States. Global development required a truly international evaluation system and this in turn meant that *ad hoc* schemes had to be abandoned and a single approach with compelling logic needed to be identified.

A team was set up with representation from marketing and clinical development from both European headquarters and the United States. Because many of the more technical aspects of a system for project prioritization have to do with probability and other related quantitative methods, a statistician was included in the team.

As implied above, the team was presented with two major tasks:

- the development of a method for project prioritization that could show long-term service with the company and assist its decision making activities in drug development on a permanent basis;

- initiating the implementation of this approach by ranking the products in the current portfolio and indicating potential candidates for termination.

## 12.3 Background from the decision consulting firm's perspective

Pharmaceutical research and development, wherein considerable sums are invested over long time horizons with low probabilities of achieving highly uncertain commercial returns, have been an active area of professional decision analysis (Balthasar *et al.*, 1978; Matheson *et al.*, 1989) for decades, since shortly after the elucidation of decision analysis principles (Raiffa, 1968; Howard, 1984).

Rather than emphasize formal principles of decision analysis, this section introduces decision analysis practice. The decision hierarchy in Fig. 12.1 illustrates the range of decision situations in which decision analysis principles have been used to guide R&D-intensive companies.

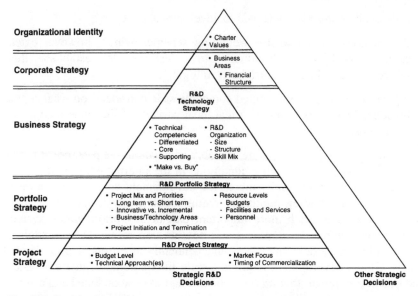

**Fig. 12.1** Decision hierarchy

At the bottom of the decision hierarchy are project-specific decisions that typically justify external decision analysis assistance only in exceptional circumstances, such as unusual, vexing choices among alternative development paths for major promising new therapies. The top of the hierarchy constitutes major strategic choices about a company's fundamental technology strategy. In the present case, Ciba's expressed need lay at the interface of project and portfolio decision making.

The 'Faster time to market' effort initiated many improvements to streamline the increasingly global development process; the project evaluation document was intended to serve as the basis for a dialogue between international project teams and decision review boards to improve decision making. The request to SDG was to evaluate all projects in the development portfolio, while training an internal group in the analytical and group facilitation skills necessary to apply the methodology in the future.

To highlight the organizational and analytical challenges implied by this need, Fig. 12.2 sets out six elements of decision quality, which provide a guide for measuring progress.

Deriving an appropriate frame requires one to choose carefully what aspects of a decision to consider. Failure to do so can result in the right

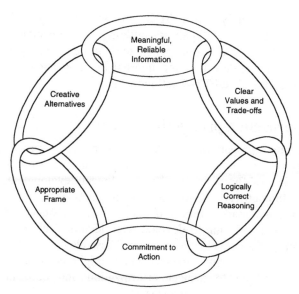

**Fig. 12.2** Elements of decision quality

answer to the wrong problem. Alternatives, information, preferences and logic are the standard elements of decision analysis theory; professional application poses considerable group facilitation challenges to effectively structure, assess and appraise a decision problem. Commitment to action requires that the appropriate people be involved from the beginning such that those responsible for ultimately deciding are ready to commit resources to the recommended action.

At Ciba, decision quality required early attention defining the projects to be evaluated (e.g. should compounds targeting multiple diseases be treated as one or multiple projects?), training in the basic principles of decision analysis, and agreement on roles and responsibilities for various individuals. The iterative interaction between the individual project teams and decision review boards makes effective use of senior management time to build quality into the effort long before the final results emerge.

SDG proposed that training be tailored to the roles of the three major groups involved: decision review boards, internal decision process leaders and content experts. Each decision board received a full-day workshop consisting of basic principles, hands-on exercises and examples before the PED team began meeting with project teams in the board's therapeutic area. Internal decision process leaders received a more extensive 2 day workshop before participating in a mini-apprenticeship consisting of an incrementally broadening role conducting project evaluations. Content experts on project teams received a minimal conceptual orientation before jumping in and learning by working through their own project.

A key element in the success of the project is that considerable internal commitment to the decision methodology existed before the project started.

The effort was initiated internally, with SDG brought in to provide leadership and to energize the start of capability building. The consultant's role was to work *through* the internal group rather than *on* the company as outsiders. The following section describes the internal thinking about the basic approach requested by the Ciba FTTM action team.

## 12.4    The basic approach: a productivity index

As part of the background preparatory work to designing the project assessment system, members of the PED team carried out interviews with employees in a wide variety of functions: marketing, clinical development, drug regulatory affairs, biometrics, research and so forth. It came as no surprise to find that eventual market return if successful, probability of success, cost of development and time to develop were identified as being the most important features of any evaluation system. These elements are of course the building blocks of decision analysis and it might be thought, therefore, that there is nothing to project prioritization and portfolio management but the implementation of such a system.

However, there are various problems to using a full decision-theoretic approach. One is that the choices that such an approach takes into account are not choices between projects, but choices between portfolios of projects. This distinction has to do with constraints on resources available. If infinite resources were available, a simple (in principle) valuation could be made for each project: is its expected net present value positive or not? If so, develop it; otherwise, abandon it. (Of course, even this simplification hides many difficulties. In particular it ignores some aspects of the option value of projects.)

The situation for most pharmaceutical companies, however, is that resources are not available to develop all projects with positive expected net present value without a major shift in financing strategy. Hence a full solution to the portfolio problem consists of examining all feasible portfolios and choosing that whose expected net present value is a maximum. Given $n$ projects with positive expected net present values, however, there are $2^n$ possible portfolios, and for even moderate $n$ this becomes far too large a number to examine exhaustively. Therefore, although the real problem remains that of choosing among the $2^n$ portfolios, it is attractive to try and study it in terms of choosing among the $n$ individual projects by ranking them in some way.

Many ways of ranking such projects have been looked at in the past. An admirable review is provided by Bergman and Gittins (1985). The task has usually been deemed to be one of finding some appropriate weights for combining discounted cost, probability of success, discounted eventual reward and any other factors considered relevant. A simple insight due to Pearson (1972), however, shows that any such weighting system must be methodologically false, whatever weights are used. If two projects had identical overall costs, rewards and probabilities of success, they would have to be ranked identically by any index based on weights *whatever the weighting system used*. Yet if one of them will fail early if it fails, whereas the other will fail late if it fails, then, other things being equal, the former must be more attractive.

To amplify this point, consider two projects. In each, the cost of a full development, which will take 6 years, will be $50m. The probability of success in each case is 0.2 and the eventual discounted reward if successful is $500m. Now, it doesn't matter what formula (e.g. multiplicative, additive, weighted) is used to rank them. If the raw inputs are the total figures given above, the projects must score identically. Supposing, however, it is learned that the question mark over the first is that it is unlikely to pass toxicological tests at the beginning. The second, however, is expected to sail through these but is unlikely to prove efficacious in phase III clinical trials. Then the bet which has to be laid in the first case to have a chance of the reward is not $50m but that relatively small fraction of $50m that is needed to determine whether the project is successful. In the second case the full $50m must be risked.

A simple game of chance may clarify this yet further. Suppose that we are offered the choice of two games. In each we must pay $3 to see the game played to completion. In each a fair coin is tossed three times and we lose unless a head shows on each occasion. In the first game the reward if successful is $25; in the second it is $20. However, in the first we must pay the $3 up front, whereas in the second, we pay $1 each time the coin is tossed and we may call a halt to the game and start again any time we like. Table 12.1 summarizes the game in terms of costs, probability of success, and rewards.

**Table 12.1  Game summary**

|  | Game | |
|---|---|---|
|  | A | B |
| Probability of success | 1/8 | 1/8 |
| Stake | $3 | $3 |
| Prize | $25 | $20 |

Put like this it appears as if the first game is the more interesting. In fact, the second is far more valuable. The expected (i.e. probability-weighted) prize for game A is $1/8 \times \$25 = \$3.125$. The expected stake is simply the total stake of $3. Hence the expected return is $\$3.125 - \$3 = \$0.125$. For game B, the expected reward is $1/8 \times \$20 = \$2.5$. However, for this game, the expected stake is not the same as the total stake. The first dollar will be paid whatever happens. However, the second will only be paid if the first toss shows a head. The third dollar will only be paid if the first two tosses each show a head. Hence, the expected stake is $1 + (1/2 \times \$1) + (1/2 \times 1/2 \times \$1) = \$1.75$. The expected return on the game is thus $\$2.5 - \$1.75 = \$0.75$. As a consequence, the expected return on game B is six times that for game A.

Even this understates the superiority of game B. Both games have positive returns. Therefore, it would be nice to have the opportunity of playing both. Neither guarantees that we will win. Therefore it is possible that we will run out of finance. But for any limit on funds available the number of losing games we can sustain cannot be fewer for game B than for game A. For example, if we have $5 available, we will not be able to play game A more than once unless we win the first time. However, we may be able to play game B three times. If the reward is expressed as a ratio of the expected stake for the two games, we have $\$0.125/\$3 = 0.042$ for game A and $\$0.75/\$1.75 = 0.429$ for game B. Hence the return on investment is 10 times as high for B as for A.

This example illustrates the essential features of the Pearson Index. In general, we assume that a project has $k$ stages. The probability of passing stage $i$, $i = 1$ to $k$, given that the project has survived to stage $i - 1$ is $p_i$. We also let $p_0$ be 1 by definition. The cost of stage $i$ is $c_i$, and the eventual return if successful is $r$. Both costs and rewards are assumed already discounted. The Pearson Index (PI) may then be expressed as

$$PI = \frac{p_1 p_2 \cdots p_k r - (c_1 + p_1 c_2 + p_1 p_2 c_3 + \ldots + p_1 p_2 \cdots p_{k-1} c_k)}{(c_1 + p_1 c_2 + p_1 p_2 c_3 + \ldots + p_1 p_2 \cdots p_{k-1} c_k)}$$

or

$$PI = \frac{r \prod_{i=1}^{k} p_i - \sum_{i=1}^{k} c_i \prod_{j=0}^{i-1} p_j}{\sum_{i=1}^{k} c_i \prod_{j=0}^{i-1} p_j} \tag{12.1}$$

Bergman and Gittins (1985) have provided a useful analogy to the portfolio problem. They suggest that it may be likened to that of packing a number of objects of differing value and size into a suitcase such that the total value of the suitcase is as large as possible. Intuitively, a reasonable solution will usually be found by ranking objects in order of the ratio of their value to volume and then packing them in that order until the suitcase is full. The reason that this does not necessarily lead to the optimal solution is that, because of edge effects, it will rarely be the case that the suitcase is exactly filled. The spare space, together with that taken up by one object already included, might be just sufficient to include two alternative objects whose value when taken together may exceed that already included. Nevertheless, packing objects in order of value to volume ratio will usually provide a good solution.

The formula given by (12.1) may also be expressed as

$$PI = \frac{r \prod\limits_{i=1}^{k} p_i}{\sum\limits_{i=1}^{k} c_i \prod\limits_{j=0}^{i-1} p_j} - 1 \tag{12.2}$$

It is also clear from (12.1) that a project is not worth developing at all if its Pearson index is not greater than 0, for this implies that the top line of the index is negative or zero and that the project is not a rational bet, whatever funding is available. Consideration of eq. (12.2) would suggest an alternative form of the index,

$$PI = \frac{r \prod\limits_{i=1}^{k} p_i}{\sum\limits_{i=1}^{k} c_i \prod\limits_{j=0}^{i-1} p_j} \tag{12.3}$$

Clearly, any ranking according to eq. (12.1) must be equivalent to any ranking according to eq. (12.3) and all that is necessary for the alternative form of the index in (12.3) is to compare it to a benchmark of 1 rather than 0 to establish whether a project is worth developing at all.

Actually, this is slightly misleading, and the form defined by eq. (12.1) is preferable. It is absolutely essential in calculating the top line of the index that the result is net of *all costs* downstream of the point at which the decision is being made. (Sunk costs should be ignored, of course.) It is not essential, however, that these costs should all appear in the denominator.

The real problem is one of optimization subject to multiple constraints. The top line of the index takes the role of an objective function. Division by expected costs is a device which deals with a given constraint: a budgetary constraint. It is not always clear, however, what the exact nature of this constraint is or should be. One might, for example, use expected development costs, or the total of the expected development and of marketing costs. Or one might use something completely different, such as, for example, expected number of staff involved.

The choice really depends on which of a number of constraints is seen as being the most pertinent. In fact, such indices in general are not dimensionless numbers but depend on the units in which both reward and 'costs' are measured. For this reason, amongst others, the PED team chose the first general form of the index. For the denominator, the expected development costs (i.e. not including marketing costs) were chosen.

## 12.5 Matching assessments and expertise: an influence diagram template

This section expands on the basic multistage problem structure introduced in Section 12.4 by describing the model templates designed to bridge the gap between the high level information needed to determine the productivity index and the narrower areas of expertise of individual members of the drug development project teams.

As a representation language for decision situations, influence diagrams serve as both an effective group communication tool and a concise mathematical model. To describe the approach taken with drug development project teams, we emphasize the influence diagram (ID) as communication tool, relying on the ample literature (Howard and Matheson, 1984; Shachter, 1986; Lauritzen and Speigelhalter, 1988) for those with greater interest in the analytical details.

A simple ID (Fig. 12.3) displays the essence of the drug development challenge and also illustrates the main features of IDs. The octagon

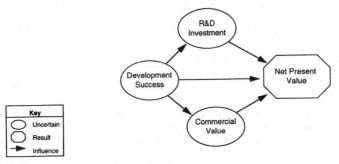

**Fig. 12.3** Core drug development influence diagram

represents the overall value measure, in this case the net present value of cash flows. Value is determined by the nodes whose arrows lead into the value node. The rectangle represents the funding decision faced by the company regarding a candidate compound. The three ovals represent uncertainties: the probability of development success, the cost of development and the commercial potential given development success.

If funded, the project's net present value is calculated by multiplying the probability of success by the net present value of commercial potential given success and subtracting from the result the expected (probability-weighted) development cost. Referring back to the previous section (eq. 12.1), the Pearson index, or net productivity index (NPI) as we shall call it henceforth, is simply the overall net present value divided by the expected development cost.

If the three summary uncertainties in Fig. 12.3 could be directly and reliably estimated, then the project evaluation would be straightforward. However, no individual possesses the expertise to internalize the complexities of drug development at this level and provide direct estimates. At the very least, such direct estimates provide precious little structured explanation of their basis or understanding of how the estimates should be updated given new information, which arrives on a regular basis as development activities proceed.

To match assessments with individual expertise and to generate greater insight into the drug development process for a candidate compound, we extend the initial influence diagram template. The first extension (Fig. 12.4)

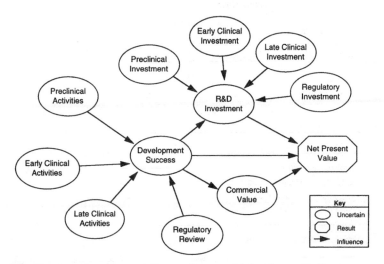

**Fig. 12.4** Drug development influence diagram with development milestones

separates the development uncertainty into key development stages and their associated investments. Although many uncertainties are resolved through-out the development process, Ciba considers four key stages for funding approval: laboratory and animal testing (pre-clinical), initial human testing (early clinical), full-scale human testing (late clinical), and review for marketing approval by regulatory authorities (regulatory review). The uncertainties in Fig. 12.4 leading into 'development success' characterize the success criteria and corresponding probability of success at each stage. Likewise, the uncertainties leading into 'R&D investment' represent the associated costs for each stage.

Separate consideration of different drug development stages allows us to generate insight into a key challenge for pharmaceutical companies: to fail early if they fail at all. Although everyone wants to have a high probability of development success, failing early rather than late can liberate resources for screening other compounds.

A decision tree view (Fig. 12.5) of the standard template displays the details of the drug development process more clearly than the corresponding influence diagram (Fig. 12.4). Although the decision tree does not explicitly represent a decision at the end of each stage, successful completion of the activities at each stage is necessary for continuation to the next stage. The success criteria are crafted to include necessary and sufficient milestones for further investment.

Thus far, we have expanded the basic template (Fig. 12.3) to include key chronological milestones (Figs 12.4 and 12.5). Yet success at any develop-

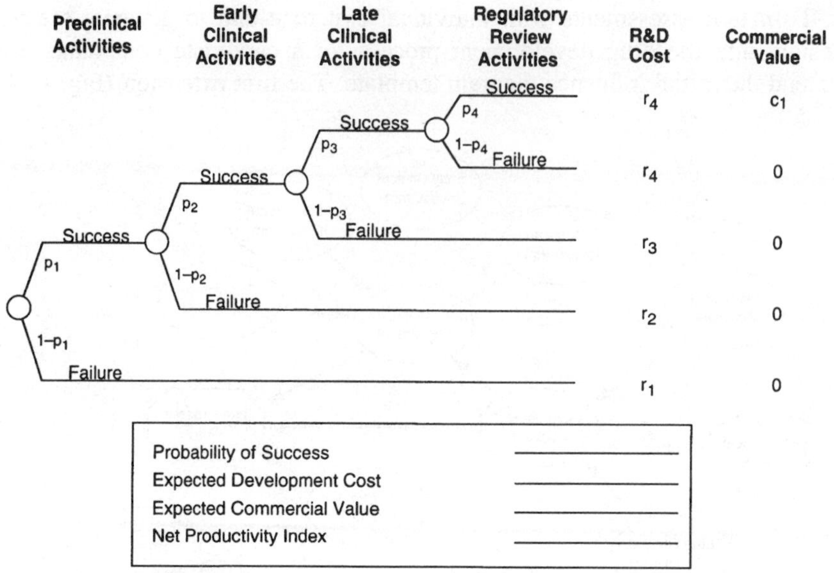

**Fig. 12.5** Drug development decision tree

ment stage requires the contribution of multiple specialities, each represented by one or more individuals on the development project team. To
match assessments to individual expertise, we extend the template once
again to capture the major areas of activity: human efficacy and safety,
animal efficacy and safety, active ingredient (the underlying compound) and
formulation (the final tablet, pill or other delivery form).

Before proceeding to describe an example, it is worth noting that the final
template displays considerable detail on the development uncertainties but
contains a direct commercial potential estimate. This reflects the initiation of
this effort at Ciba within the development division. The prospects for
extending the template to explicit consideration of commercial uncertainties
are discussed in Section 12.8. The process for doing so and the insights
generated thereby are described in the professional literature (Matheson *et
al.*, 1989).

## 12.6 A project evaluation full-cycle example

This section describes the project evaluation process for an illustrative
example in step-wise fashion (Table 12.2), including the 'softer' aspects of
the joint work, such as preparation, the sequence of meetings with the
development project team, sharing of facilitation responsibilities by the
client core team and the consultants, review by the decision board and
follow-up.

The steps described in Table 12.2 were developed over the first several
months of the joint collaboration using two real projects. With further
refinements along the way, and numerous customizations for special circums-

**Table 12.2** Project evaluation activities and responsibilities

- Advance meeting preparation
- Initial meeting with project team at location of team leader
- Follow-up meeting with project team at second location
- Individual follow-up for specific assessments
- Resolve development uncertainties with team leader
- Obtain R&D cost estimates
- Obtain commercial estimates
- Perform calculations
- Present results to team
- Present results to review board
- Address challenges with team

tances, the process was next applied to a therapeutic area of 10 projects, and then broadened to include the entire portfolio of development projects.

The 'PED team' of process facilitators was organized to interact with international development project teams located primarily in Switzerland and the United States. A two-person PED team, consisting of an external consultant and an internal planning organization individual, was formed in each location. Initially, most steps were led by the external consultant, with support from the internal partner. Roles reversed over time, however, as the internal PED team member became sufficiently comfortable to take over responsibility for leading project teams through the exercise. Coordination between the two PED team pairs in each location minimized the travel required for development project team members.

We now elaborate on each step.

### 12.6.1   Advance meeting preparation

The partners in the two-person PED teams bring complementary knowledge to the process. The internal partner tends to know or have access to background information on the nature of the compound under development by the project team. Perhaps more importantly, the internal partner knows the group dynamics of the project teams and the likely sticking points of applying the general form of the PED methodology.

The external partner knows the methodology and can suggest various ways that particular project issues might be handled. Consideration of special issues in advance makes for a smoother meeting interaction with the project team. Logistics and sharing of meeting facilitation and note-taking roles must also be agreed to in advance to ensure a smooth team performance.

### 12.6.2   Initial meeting with project team at location of team leader

As mentioned earlier, most development project teams are international, with most members located in the United States or Switzerland. By conscious design, the first PED meeting takes place at the location of the international project team leader. After a brief overview of the motivation behind the PED effort and the structure of the PED influence diagram template, the session turns to the structure of the project.

Starting with the current stage of development (e.g. pre-clinical, early clinical or late clinical), the facilitator leads the project team to clarify the success criteria for the activities currently underway. Taking one hurdle (e.g. human efficacy) at a time, the team, represented primarily by one or several experts, articulates as clearly as possible the minimum level of success that will justify continuing on to the next stage of development. Note that this is not the team's best guess at what will happen but the minimum level of success required to continue.

In many instances, such success criteria had already been determined through previous project team activities and the PED leaders simply recorded the result. Quite often, however, members of the team would challenge the success criteria and a lively debate would ensue: 'How much efficacy do we need in early clinical trials?'; 'Would we really stop the project if we obtained a certain toxicology finding?'; 'I know that the regulatory authorities don't require a certain safety level, but are *we* ready to commit resources for large-scale trials without more evidence?'. The most valuable discussions occurred when members of the project team began to share information freely across their specialities, and to understand more deeply the issues that others were handling.

Only after clearly specifying the success criteria for a given hurdle would the conversation move to assessing the probability of achieving success. This is the least familiar aspect of the methodology to the project teams. To obtain the widest possible range of opinions with the least bias, the team members are invited to write down their assessments individually. In many cases, only experts express an opinion. Non-experts, however, are encouraged to express their opinions as well. Next, the range of estimates are collected and shared with the group.

When little disagreement exists or a single expert makes a non-controversial assessment, the meeting moves on. Otherwise, advocates of extreme positions are asked to explain their reasoning and the facilitator leads the team through an exploration of different viewpoints. Ultimately, the team reaches some closure, either with a consensus probability assessment or a set of questions that must be addressed by a subgroup. The power of the approach is that, fairly often, only when judgements are expressed in probabilistic terms do team members recognize how significantly different their viewpoints actually are.

Generally, this first session runs for 2–5 hours, depending on the complexity of the project. The vast majority of the discussion tends to focus on the two or three difficult issues the project team faces in designing or interpreting the early clinical activities. The most important output of the meeting is greater team alignment on the key issues and how the different hurdles relate to each other.

### 12.6.3   Follow-up meeting with project team at second location

The PED team pair that ran the first meeting must then organize the results of the meeting for use by their counterparts in the other location. A structured template for recording the hurdle success criteria and probability assessments was developed over time to streamline the communication process. Often, telephone conversations are necessary to discuss specific issues.

A key strength of the PED process is that it surfaces differences of opinion between the local teams in each geographic region. Participants in the second meeting are asked to review and challenge the results of the first meeting. Overlooked issues are often raised and questions recorded about success criteria or probability assessments. Importantly, the participants are allowed to craft different success criteria and provide their own probability assessments, but they cannot overrule the results of the first meeting, whose participants are not present. Differences of opinion must ultimately be resolved by the project team leader.

### 12.6.4   Individual follow-up for specific assessments

One or more issues are often raised in the first two meetings that cannot be answered at the time, or key individuals are not able to attend. The PED team must therefore convene individuals or subgroups to provide success criteria and probability assessments.

### 12.6.5   Resolve development uncertainties with team leader

By this time, the PED team is able to put together a package consisting of the success criteria for the various hurdles and the PED influence diagram, putting all the pieces together in probabilistic terms. The international project team leader bears the responsibility to resolve differences of opinion so that the team ultimately expresses a single view of their project to the rest of the organization.

### 12.6.6   Obtain R&D cost estimates

From the project team's development plan, a standardized procedure is used to estimate costs for each stage of development. Special costs such as facility investments and licensing milestone payments are handled on a case-by-case basis.

### 12.6.7   Obtain commercial estimates

Based on the minimum success criteria articulated by the development project team, the marketing department generates an estimate of the commercial potential for the product. Currently, commercial estimates are single scenario forecasts. The scenario represents the best estimate of commercial potential, given that the minimum success criteria are achieved. Note that this best estimate is generally considerably better than the minimum level of success. See Section 12.8 for further comments on treatment of uncertainty in the commercial estimates.

### 12.6.8   Perform calculations

The development probabilities for individual hurdles, R&D costs by stage and commercial estimate given success are entered into an influence diagram model to produce the PED summary tree and supporting documents. Note that, of the 15 or so hours of conversations that have gone into the project so far, the computer-based modelling and calculations typically require less than 15 min. For most projects, the primary benefits are simplicity, transparency and team alignment, rather than analytical sophistication.

Nonetheless, some projects have critical interproject dependencies that cannot be handled appropriately by the standard template. These projects are treated separately by extending the standard template to incorporate the special issues. The power of the decision-analytic approach is that such dependencies can be incorporated in a straightforward, consistent manner. The PED team is responsible for judging whether the added complexity is truly of benefit to the project team.

### 12.6.9   Present results to the international project team

Before presentation to the review board that is responsible for funding recommendations, the project team has an opportunity to review the results of the evaluation integrating their success criteria and probabilities, the R&D costs and the marketing estimates. When possible, these meetings are held in one location or by videoconference so that the entire project team can understand, and challenge where appropriate, the evaluation results.

### 12.6.10   Present results to review board

Ciba's development portfolio is organized into four therapeutic areas, each having its own review board consisting of representatives from each development line function. The therapeutic area review board PED meeting consists of presenting each project in a similar format, providing the opportunity to discuss and challenge individual components of the evaluation. The PED process serves as a communication channel between the review board and the project teams. The review board can challenge success criteria or probability assessments based on individual project knowledge or calibration judgements across projects in the therapeutic area. A separate process beyond the scope of this paper is used to achieve portfolio calibration across therapeutic areas.

### 12.6.11   Address challenges with the project team

Challenges identified in the review board meeting must be followed up with the project teams. In some cases, the project team leader may be invited to a future review board meeting to address a specific issue. Alternatively, the specific challenges may be addressed within the project team and the updated evaluation forwarded to appropriate individuals.

Beyond the first iteration with the PED methodology, updates are made when significant new information becomes available. The PED becomes a focal point for communicating changes in project status within the organization. Use of the individual project evaluations to generate insights into each therapeutic area portfolio and the overall development portfolio is beyond the scope of this article.

## 12.7   Lessons learned

The collaboration between Ciba and SDG has been a tremendous learning experience for all parties. The joint implementation approach has made the most of Ciba's organizational readiness to improve its project evaluation process and SDG's methodological depth, and the capability-building commitment of both parties.

Strengths of the efforts to date include the following:

- Active support of senior management proved essential and gave the project a good start.

- The internal staff and external consultants established a good working relationship, including mutual recognition of client requirements and consultant capabilities.

- The objective of capability transfer from consultants to a specific cross-functional group at the client increases the sense that this is a Ciba effort and not something outsiders are doing *to* Ciba. Classroom and hands-on training are supporting capability transfer.

- Modelling software has been customized to reflect Ciba-specific templates.

- Many development project teams recognized that deeper understanding of their own projects resulted from the process.

- A number of concrete changes in project plans have resulted from the effort.

- Level and quality of debate at decision review boards have increased.

Challenging hurdles encountered to date that must be overcome for continued success include the following:

- Mistakes of any kind in implementing or presenting a new approach can discredit the methodology rather than the input.

- Developing group facilitation skills is critical for capability transfer; technical skill development alone is insufficient.

- Acceptance is limited for some by lack of confidence in calibration of assessments (i.e. consistency across projects). Multiple cross-project reviews have been established to address this challenge.

- A data base storage system is needed to track assessments over time (who said what when) and support peer review within projects and within functional units (e.g. toxicology) across projects.

- Communication logistics difficulties hamper consistent application of the methodology in Europe and the US; such challenges will increase with extension soon to Japan.

## 12.8   Future challenges

We briefly list here some challenges and opportunities that are outside the scope of our current effort but are natural extensions for the future:

- Explicitly characterize uncertainty in commercial assessments and improve integration of development and marketing organizations. Increased transparency and insight into uncertainty in development are raising interest to provide the same for marketing estimates.

- Explicitly treat uncertainty in development investment assessments. While less significant perhaps than commercial uncertainty, development cost and timing uncertainty are critical to overall productivity.

- For all assessments, the more we rely on averages (e.g. industry benchmarks), the less we can distinguish between projects.

- We anticipate deeper training needs as requests move from the current standard template to unique, complex, high-stakes single decisions.

- Project evaluation provides insight into portfolio balance with respect to projects in various phases and diversification across therapeutic areas.

- Option theory provides insight into the importance of timing and risk aversion for project evaluation.

- The probabilistic view of project evaluation may yield insight into constraints (e.g. toxicology staff) other than the financial implications highlighted by the net productivity index.

- The methodology may be adapted to handle licensing candidates.

- The underlying principles of transparency and consistency may be applied to assist with evaluating research candidates, which do not have clear clinical plans or commercial targets.

## References

Balthasar, H.U., Boschi, R.A.A. and Menke, M.M. 1978: Calling the shots in R&D. *Harvard Business Review*, May–June.

Bergman, S. and Gittins, J. 1985: *Statistical methods for pharmaceutical research planning*. New York: Dekker.

Howard, R.A. 1984: Decision analysis: applied decision theory. In Howard, R.A. and Matheson, J.E. (eds) *Readings on the principles and applications of decision analysis*. Menlo Park, CA: Strategic Decisions Group, vol. 1, 97–113.

Howard, R.A. and Matheson, J.E. 1984: Influence diagrams. In Howard, R.A. and Matheson, J.E. (eds) *Readings on the principles and applications of decision analysis*. Menlo Park, CA: Strategic Decisions Group, vol. 2, 721–62.

Lauritzen, S.L. and Spiegelhalter, D.J. 1988: Local computations with probabilities on graphical structures and their application to expert systems. *Journal of the Royal Statistical Society* **50**, 157–224.

Matheson, J.E., Menke, M.M. and Derby, S.L. 1989: Managing R&D portfolios for improved profitability and productivity. *The Journal of Science Policy and Research Management* **4**, 400–12.

Pearson, A.W. 1972: The use of ranking formulae in R&D projects. *R&D Management* **2**, 69–73.

Raiffa, H. 1968: *Decision analysis: introductory lectures on choice under uncertainty*. Reading, MA: Addison-Wesley.

Shachter, R.D. 1986: Evaluating influence diagrams. *Operations Research* **34**, 871–82.

# Index